Jones, P (Paul)

GEL ELECTROPHORESIS: NUCLEIC ACIDS

ESSENTIAL TECHNIQUES SERIES

Series Editor
D. Rickwood
Department of Biological and Chemical Sciences, University of Essex, Wivenhoe Park, Colchester, UK

Published titles
Antibody Applications
Gel Electrophoresis: Nucleic Acids

Forthcoming titles
DNA Isolation and Sequencing
PCR
Gel Electrophoresis: Proteins
Gene Transcription
Human Chromosome Preparation
Cell Biology

GEL ELECTROPHORESIS: NUCLEIC ACIDS

ESSENTIAL TECHNIQUES

P. Jones

Department of Biological and Chemical Sciences, University of Essex, Colchester, UK

JOHN WILEY & SONS

Chichester • New York • Brisbane • Toronto • Singapore

Published in association with BIOS Scientific Publishers Limited

© BIOS Scientific Publishers Limited, 1995. Published by John Wiley & Sons Ltd, Baffins Lane, Chichester, West Sussex PO19 1UD, UK, in association with BIOS Scientific Publishers Ltd, 9 Newtec Place, Magdalen Road, Oxford OX4 1RE, UK.

British Library Cataloguing in Publication Data
A catalogue record for this book is available from the British Library.

ISBN 0 471 96043 8

Library of Congress Cataloging in Publication Data
Jones, P. (Paul)
    Gel electrophoresis : nucleic acids / P. Jones.
       p. cm. ––(Essential techniques)
    Includes bibliographical references and index.
    ISBN 0–4–716043–8 (alk. paper)
    1. Nucleic acids––Analysis. 2. Gel electrophoresis. I. Title.
    II. Series : Essential techniques series.
QP620.J66   1999
574.87′328––dc20   95–25016
                                                                          CIP

Typeset by Footnote Graphics Ltd, Warminster, UK
Printed and bound in UK by Biddles Ltd, Guildford, UK

The information contained within this book was obtained by BIOS Scientific Publishers Limited from sources believed to be reliable. However, while every effort has been made to ensure its accuracy, no responsibility for loss or injury occasioned to any person acting or refraining from action as a result of the information contained herein can be accepted by the publishers, authors or editors.

CONTENTS

Contents

ABBREVIATIONS

A	adenine
A_{260}	absorbance at 260 nm
% acrylamide	polyacrylamide gel concentration expressed in terms of total monomer (i.e. acrylamide and cross-linker)
AMPPD	disodium 3-(4-methoxyspiro[1,2-dioxethane-3,2′-tricyclo-[3.3.1.1] decan]4-yl) phenylphosphate
APS	ammonium persulfate
ATA	aurintricarboxylic acid
ATP	adenosine triphosphate
BAC	*N,N′-bis*-acrylylcystamine
Bisacrylamide	*N,N′*-methylene-*bis*-acrylamide
bp	base pair
Bq	Becquerel (1 disintegration/sec)
BSA	bovine serum albumin
C	cytosine
CAIP	calf alkaline intestinal phosphatase
Ci	Curie (3.7×10^{10} Bq)
cDNA	complementary DNA
c.p.m.	counts per minute
CSPD	5-chloro derivative of AMPPD
CTP	cytosine triphosphate
Da	dalton
DADT	diallyltartardiamide
DEPC	diethylpyrocarbonate
DMSO	dimethylsulfoxide
DNase	deoxyribonuclease
ddNTP	dideoxyribonucleotide
dNTP	deoxyribonucleotide
dsDNA	double-stranded DNA
dsRNA	double-stranded RNA
DTT	dithiothreitol
ECL	enhanced chemiluminescence
EDTA	ethylenediaminetetra-acetic acid (disodium salt)
g	relative centrifugal force
G	guanine
HEPES	4-(2-hydroxyethyl)piperazine-1-ethanesulfonic acid
HRP	horseradish peroxidase

IAA	isoamyl alcohol		rNTP	ribonucleoside triphosphate
			r.p.m.	revolutions per minute
kb	kilobase		rRNA	ribosomal RNA
kbp	kilobase pair		RT	room temperature
kBq	kilobecquerel			
			SDS	sodium dodecyl sulfate/sodium lauryl sulfate
LB agar/broth	Luria-Bertani agar/broth		SSC	saline sodium citrate
			ssDNA	single-stranded DNA
M	molarity		ssRNA	single-stranded RNA
MES	2-morpholinoethanesulfonic acid			
MOPS	3-morpholinopropanesulfonic acid		T	thymine
mRNA	messenger RNA		TE	10 mM Tris-HCl pH 8.0, 1mM EDTA
			TAE	Tris-acetate-EDTA electrophoresis buffer
nt	nucleotide		TBE	Tris-borate-EDTA electrophoresis buffer
			TBS	Tris-buffered saline
PBS	phosphate-buffered saline		TEMED	N,N,N',N'-tetramethylethylenediamine
PCR	polymerase chain reaction		TGGE	temperature gradient gel electrophoresis
pH	hydrogen ion exponent		T_m	melting temperature
PIPES	piperazine-1,4-bis(2-ethanesulfonic acid)		Tris	Tris(hydroxymethyl)aminomethane
PMSF	phenylmethylsulfonyl fluoride		tRNA	transfer RNA
Poly(A)$^+$ RNA	polyadenylated RNA			
PVS	polyvinyl sulfate		U	uracil
			UV	ultraviolet
RFLP	restriction fragment length polymorphism			
RNase	ribonuclease			

PREFACE

A wide range of electrophoresis techniques have been developed for the separation and analysis of nucleic acids and nucleic acid:protein interactions. Electrophoresis has become an indispensable tool for the research of cell and molecular biologists because it is a simple, rapid and highly sensitive technique for both preparative and analytical separations.

This volume of the *Essential Techniques* series covers procedures required for carrying out electrophoretic and molecular analyses of nucleic acids. This book seeks to provide key techniques for the electrophoresis of RNA and DNA, and their subsequent analysis and recovery. Techniques involving analysis of nucleic acid:protein interactions have also been described. I hope that the reader will find this book to be instructive and useful in their work on a daily basis.

Paul Jones

SAFETY

Attention to safety aspects is an integral part of all laboratory procedures and national legislations impose legal requirements on those persons planning or carrying out such procedures. While the author, editor, and publisher believe that the recipes and practical procedures, as set forth in this book, are in accord with current recommendations and practice at the time of publication, they accept no legal responsibility for any errors or omissions, and make no warranty, expressed or implied, with respect to material contained herein. It remains the responsibility of the reader to ensure that the procedures which are followed are carried out in a safe manner and that all necessary safety instructions and national regulations are implemented.

In view of ongoing research, equipment modifications and changes in governmental regulations, the reader is urged to review and evaluate the information provided by the manufacturer, for each reagent, piece of equipment or device, for any changes in the instructions or usage and for added warnings and precautions.

These are freely accessible using:
http://joule.pcl.ox.ac.uk/MSDS/.

Other safety information on the Internet can be accessed on:
gopher://atlas.chem.utah.edu/11/MSDS
gopher://ginfo.cs.fit.edu:70/lm/safety/msds
http://physchem.ox.ac.uk/MSDS
http://www.fisher1.com/Fischer/Alphabetical Index.html
http://www.pp.orst.edu

You are actively encouraged to check these data sheets to confirm our assignments and for more detailed information on individual hazards; however the author, editor and publisher can accept no responsibility for any material contained in these data sheets. Furthermore, you must always follow the precautions outlined on labels and data sheets provided by individual manufacturers.

Specific chemical hazards
Acrylamide: toxic in powder and liquid form. Wear a face mask to avoid inhaling powder when preparing solutions and wear gloves when making and handling gels.

All procedures mentioned within this book must be carried out under conditions of good laboratory practice in accordance with local and national guidelines. Some procedures involve specific hazards, including but not limited to hazards in the following categories:

Chemical. A number of the reagents are known to be carcinogenic, mutagenic, toxic, inflammable, highly reactive or otherwise hazardous. Substances known to be hazardous have been marked with the symbol △ in the list of reagents (but not subsequently) for each protocol, or if they appear as alternatives to the main protocol, the *first time* they appear in the notes. The reader should consult the safety notes on these pages before embarking on any of the procedures covered. This is in no way meant to imply that undesignated chemicals are nonhazardous, and all laboratory chemicals should be handled with extreme caution. Information is not available on the possible hazards of many compounds. The criteria we have generally used for denoting a substance with △ is based upon a hazard level of 2 or more (on a scale 0–4) in any of the categories in the Baker Saf-T-Data™ system used in the material safety data sheets (MSDS) held at the University of Oxford, UK.

Formaldehyde: toxic in contact with skin and forms a choking vapor and so must be handled in a fume cupboard. Wear gloves and goggles at all stages of experimentation with this chemical.

Phenol: a suspected carcinogen and as such should be used in fume cupboards. Phenol burns skin very badly; volumes of 10 ml can lead to serious medical problems. If burns occur despite protection, the burnt area should be washed with 20% (w/v) polyethylene glycol (PEG) in 50% (v/v) industrial methylated spirits (IMS) **NOT** water. At least a liter of this solution should be immediately available at all times.

Radiation. The use of radioisotopes is subject to legislation and requires permission in most countries. Furthermore, national guidelines for their use and disposal must be rigorously adhered to. The procedures in protocols that use radioisotopes must only be carried out by individuals who have received training in the use of such material using the appropriate facilities, protection and personal monitoring procedures.

Biological. Antibodies, sera and cells (particularly, but not exclusively, those of human and nonhuman primate origin) pose a significant biological hazard. All such materials, whatever their origin, may harbor human pathogens and should be handled as potentially infectious material in accordance with local guidelines. Any recombinant DNA work associated with protocols is likely to require permission from the relevant regulatory body and you must consult your local safety officer before embarking upon this work.

Electrical. Many of the procedures in this book use electrical equipment. Electrophoresis techniques may present particular hazards of this nature.

Lasers. Flow cytometers and certain other types of laboratory equipment contain lasers. Users should ensure they are fully aware of the potential hazards of using such equipment.

I   INTRODUCTION TO GEL ELECTROPHORESIS

Electrophoresis is a technique that uses the migration of molecules in an electrical field to obtain separations. The rate of migration depends on the strength of the electrical field, together with the net charge, size and conformation of the molecules, and on the ionic strength, viscosity and temperature of the media through which the molecules are moving.

Properties of nucleic acids and nucleoproteins

Nucleic acids are made up of nucleotides. A nucleotide consists of a five carbon sugar, a purine/pyrimidine base and a phosphate group. The five bases, adenine (A), cytosine (C), guanine (G), thymine (T) and uracil (U) are the major bases of DNA (A, G, C, T) and RNA (A, G, C, U + modified bases). In both DNA and RNA, nucleotides are joined via a $3'$–$5'$ phosphodiester linkage of the phosphate group of one nucleotide to a hydroxyl group on the sugar of the adjacent nucleotide. Structurally, DNA is usually two complementary, antiparallel strands, whilst RNA is usually a single-stranded molecule but virtually always forms helical secondary structures. The phosphate group of each nucleotide contributes a net charge of −1 per nucleotide to a chain of nucleotides and forms the basis of electrophoretic separations. The majority of nucleic acids are found associated with proteins *in vivo* forming defined nucleoprotein complexes and these can be also separated by electrophoresis either with or without fixation of the complex.

Nuclease inhibitors

Nuclease inhibitors are included in gel electrophoresis procedures to prevent nucleic acid degradation during electrophoresis. *Table 1* lists the main nuclease inhibitors used in many laboratory procedures. DEPC-treated water used to make up all gel solutions should be autoclaved prior to use to prevent DEPC interfering with any further analyses. Inhibitors are also often used during the isolation of RNA and DNA to prevent degradation by the nucleases present in cells.

General manipulation procedures should also be followed to prevent nuclease action: (i) wear disposable plastic surgical gloves during all manipulations, (ii) autoclave or filter-sterilize all solutions, (iii) wipe surfaces clean prior to starting experiments, (iv) avoid leaning over or breathing into samples, and (v) autoclave or bake at 200°C all ceramics and glassware prior to experiments.

Table 1. Nuclease inhibitors and their working concentrations

Agent	Active concentration	Method of nuclease inactivation	Comments
Aurintricarboxylic acid (ATA)	10 μM	Complexes to a wide range of nucleases	
Bentonite	3 mg/ml	Inactivation by adsorbing to nucleases	
2-Mercaptoethanol	0.1–0.25 M	Reduces disulfide bonds, denaturing proteins	Very toxic
Diethylpyrocarbonate (DEPC)	0.1% (v/v)	Alkylates proteins, disrupting protein structure	Toxic
Dithiothreitol (DTT)	1 mM	Reduces disulfide bonds, denaturing proteins	
Ethylenediaminetetra-acetic acid (EDTA)	1–10 mM	Chelates divalent cations needed for ribonuclease activity	
Guanidine hydrochloride	8 M	Inactivates ribonucleases	Toxic
Guanidium thiocyanate	4 M	Inactivates ribonucleases	Toxic, strongest agent for ribonuclease inactivation
Heparin	0.5 mg/ml	Binds to basic ribonucleases	
8-Hydroxyquinoline[a]	0.1% (w/v)	Inactivates ribonucleases	Very toxic
Macaloid	0.015% (w/v)	Adsorbs to ribonucleases	
Phenol/Chloroform	50% (v/v)	Denatures ribonucleases	Toxic
Polyvinyl sulfate (PVS)	1–10 μg/ml	Complexes to basic nucleases	
Proteinase K	100–200 μg/ml	Hydrolysis of proteins	
Sodium dodecyl sulfate (SDS)	0.1–1% (w/v)	Disrupts protein structures	
Ribonucleoside vanadyl complex	10 mM	Binds to active site of ribonuclease	

[a] Should be added to phenol prior to deproteinization.

II ISOLATION OF DNA AND RNA

The most common methods of DNA and RNA extraction are based around cell disruption followed by phenolic removal of proteins and nucleic acid preciptation. The procedures used for cell disruption vary with tissue type. Bacteria are usually lysed in a sodium dodecyl sulfate (SDS) lysis buffer before being extracted with a phenol procedure. Animal cells can be taken from tissue culture or directly from a donor. The former source allows easy collection of cells, normally by scraping a rubber policeman over the sheet-like layers in the culture dishes, pouring media off and lysing the cells directly. Tissue taken directly from a donor source is not often fluid and so must be broken apart as well as lysed. The usual method of breaking tissue apart is homogenization which can be achieved by hand grinding with a mortar and pestle, by use of a Polytron blender or homogenizing vessel. This procedure can be enhanced for tough tissues by the use of proteolytic enzymes such as collagenase. Plant cells have a cellulose cell wall surrounding their cell membrane which must be broken by strong mechanical action. This is usually achieved by grinding fresh or frozen tissue in prechilled mortars in a pool of liquid nitrogen until a fine powder is achieved, before transferring to a deproteinization solution where further cell lysis is accomplished by high speed shearing with a Polytron blender.

A few golden rules can be given and it is important that they are followed if the isolation procedure is to be successful; all too often the first attempts to isolate DNA or RNA leave the novice with only degraded fragments. Firstly, it is important to ensure that all of the reagents used are as free as possible of nucleases. All solutions should be made from fresh untouched solids with autoclaved double-distilled water and autoclaved where possible. Solutions used in the initial extraction (during which nucleases are released) are often supplemented with nuclease inhibitors as described previously. A pH range of 6.5–7.0 for solutions is normally a necessity to ensure that no degradation occurs, particularly for RNA as it has a tendency to be alkali labile. Glassware should be oven baked at 200°C for 4–12 h; plastic is usually autoclaved. Skin is a notorious source of nucleases and so it is essential to wear disposable plastic surgical gloves during all manipulations.

Methods available

RNA extraction (see *Protocol 1*)

RNA extractions are more difficult than DNA extractions due to the presence of high levels of ribonucleases in cellular cytosols. The preferred method for inactivating RNases during cell disruption is by using the chaotrope guanidinium thiocyanate. The most frequently used isolation method is detailed in *Protocol 1* and other specialized isolation methods may be found in refs 1–3.

Problems: High levels of ribonuclease in tissues necessitate quick extraction. Guanidinium thiocyanate and phenol are toxic and must be handled with care.

DNA extraction (see *Protocol 2*)

Genomic DNA isolation methods are usually specific to a certain type of tissue. *Protocol 2* gives a method that provides high quality DNA from mammalian cells which can also be carried out according to procedures detailed in refs 4 and 5. Plant genomic DNA can be isolated according to *Protocol 2* or according to ref. 6.

Problems: Genomic DNA extractions are prone to shearing of the very long DNA molecules during isolation.

Preparation of Sepharose CL6B spin columns (see *Protocol 3*)

Column chromatography using separation media such as Sepharose CL6B can be used to remove small molecules (e.g. oligonucleotides/

References

1. Sambrook J., Fritsch, E.F. and Maniatis, T. (eds) (1989) *Molecular Cloning – A Laboratory Manual*, 2nd edn. Cold Spring Harbor Laboratory Press, New York.
2. Jones, P.G., Qiu, J. and Rickwood, D. (1994) *RNA Isolation and Analysis*. BIOS Scientific Publishers, Oxford.
3. Chabot, B. (1994) in *RNA Processing – A Practical Approach*, Vol. I (Higgins, S.J. and Hames, B.D., eds). IRL Press at Oxford University Press, Oxford.
4. Kupiec, J.J., Giron, M.L., Vilette, D., Jeltsch, J.M. and Emanoil-Ravier, R. (1987) *Anal. Biochem.* **164**:53.
5. Bowtell, D.L.L. (1987) *Anal. Biochem.* **162**:463.
6. Dean, C., Sjodin, C., Page, T., Jones, J. and Lister, C. (1992) *Plant J.* **2**:69.
7. Shultz, D.J., Craig, R., Cox-Foster, D.L., Mumma, R.O. and Medford, J.I. (1994) *Plant Mol. Biol. Rep.* **12**:310–316.
8. Chomcynski, P. and Sacchi, N. (1987) *Anal. Biochem.* **162**:156.
9. Blin, N. and Stafford, D.W. (1976) *Nucl. Acids Res.* **3**: 2303.

Protocols provided

1. *Isolation of total RNA using the acid guanidinium thiocyanate method*
2. *Isolation of genomic DNA from mammalian cells*
3. *Sepharose CL6B spin columns for nucleic acid purification*
4. *Spectrophotometric measurement of nucleic acid concentration*

degraded fragments) from a DNA or RNA preparation. Always test new column media with a proportion of the sample before committing all of the preparation to column chromatography. Separose CL6B spin columns (*Protocol 3*) are routinely used to remove oligonucleotides and salts from many DNA samples prior to further molecular manipulation.

Determination of concentration and purity of nucleic acids using optical density (see *Protocol 4*)

Spectrophotometric assessment of nucleic acid concentration is based upon the strong absorption of light at 260 nm of DNA/RNA. The absorption is measured at the peak of the nucleic acid spectra and concentrations can be calculated reasonably accurately using the equations detailed below:

dsDNA:
$50/(1/OD_{260}$ of the RNA sample) = concentration of DNA sample (μg/ml)

ssDNA/RNA:
$40/(1/OD_{260}$ of the RNA sample) = concentration of ssDNA/RNA sample (μg/ml)

This equation is based on a sample of 40 μg RNA/ml producing an OD_{260} of 1.0. The advised region for optical density (OD) measurement is between an OD_{260} of 0.1 and 0.5. Many laboratories use this method of quantitation to establish nucleic acid concentration prior to further experiments. The concentration of protein is also assessed as a measure of purity of the sample. This is achieved by measuring absorbance at 280 nm. A pure sample of DNA or RNA should have a ratio of absorbance at 260/280 nm exceeding 1.75 for further analytical purposes. If the sample is impure phenol extractions should be repeated to remove contaminating proteins. Excess phenol can then be removed by extracting the aqueous phase with ether. It should be noted that recovery of DNA or RNA during any step is never 100% so each extraction will involve the loss of some nucleic acid. If the sample is opalescent then this is often an indication that polysaccharide impurities are present and protocols designed for polysaccharide-rich tissues should be used (e.g. ref. 7).

Problems: The ratio of $A_{260/280}$ for DNA and RNA to protein has a theoretical maximum of 2:1 but values in excess of this will occur if the sample contains organic solvents such as phenol or chemicals such as guanidinium thiocyanate. A_{230} values greater than A_{260} indicate organic solvent contamination (e.g. ethanol or isopropanol) or can indicate contamination by phenol or guanidium. A_{325} values greater than zero indicate particulate matter is present in the sample.

Isolation of DNA and RNA

Protocol 1. Isolation of total RNA using the acid guanidinium thiocyanate method

Reagents

Ethanol
GTC solution (4 M guanidinium thiocyanate, 25 mM sodium citrate (pH 7), 0.5% sarcosyl, 0.1 M 2-mercaptoethanol⚠
Phenol:chloroform:isoamyl alcohol (IAA) (25:24:1) ⚠
3 M Sodium acetate (pH 5.2)

Equipment

Homogenizer/Polytron blender
Microcentrifuge

Technique (ref. 8)

1 Homogenize cells in four volumes of GTC solution using a hand homogenizer ('dodel')/Potter homogenizer/Polytron blender.

2 'Phenol extract' by adding an equal volume of phenol:chloroform:IAA and shake to form an emulsion. Centrifuge for 5 min at 10 000 g to separate the phases. Transfer the upper aqueous layer to a fresh tube.①

3 'Ethanol precipitate' by adding 0.1 vol. 3 M sodium acetate and 2.5 vol. ethanol. Invert to mix and place at −20°C for 1 h. Collect the precipitate by centrifugation at 15 000 g for 10 min at 4°C. ⒈

4 Redissolve the pellet in 500 μl of sterile double-distilled water. Phenol extract the sample as in step 2 with 500 μl of phenol:chloroform:IAA.②

Notes

① Samples containing high levels of proteins should be phenol extracted until the phase interface is clear.

② Large gelatinous pellets are produced from high polysaccharide tissue which should be removed by using other procedures [7].

③ Total RNA should produce three rRNA bands above the bromophenol dye line and an mRNA smear close to the dye line. Degraded RNA is smaller and runs ahead of the dye.

5 Ethanol precipitate the RNA in the recovered aqueous phase as in step 3 with 50 μl (0.1 vol.) 3 M sodium acetate and 1.25 ml (2.5 vol.) ethanol. Invert to mix and place at −20°C for 1 h. Collect the precipitate by centrifugation at 15 000 *g* for 10 min at 4°C.[2]

6 Redissolve the pellet in 20 μl sterile double-distilled water. Run 1 μl total RNA on a 1.5% agarose gel to assess RNA integrity (*Protocol 14*).③

7 Quantitate 5 μl RNA sample in a 400 μl quartz cuvette at 260 nm according to *Protocol 4*.

Pause points

[1] RNA may be left at −20°C overnight without significant degradation.
[2] May be left at −20°C for up to 24 h.

Protocol 1. Isolation of total RNA

Reagents

10 M Ammonium acetate (filtered)
Buffer-saturated phenol (pH 8)⚠①
Ethanol⚠
70% Ethanol⚠
Lysis solution (10 mM Tris HCl (pH 8), 0.5% sodium dodecyl sulfate (SDS), 0.1 M ethylenediaminetetra-acetic acid (EDTA), 20 µg/ml pancreatic RNase)
20 mg/ml Proteinase K

TE (10 mM Tris, 1 mM EDTA)
Tris-buffered saline (TBS; 25 mM Tris base, 0.2 M NaCl, 2.5 mM KCl, 15 µg/ml phenol red (pH 7.4), autoclaved)

Equipment

Bench-top centrifuge with swing-out rotor
Erlenmeyer flask
Orbital table
Water baths at 37°C and 50°C

Technique (ref. 9)

1 Gently wash monolayers of cells twice with 5 ml ice-cold TBS. Decant TBS leaving 0.5 ml in the petri dish. Harvest the cells by scraping a rubber policeman over the sheet-like layers in the culture dishes and transfer the cells to a tube on ice. Wash the dish with 1 ml ice-cold TBS to recover any cells remaining. Add the wash to the cells on ice.②

2 Pellet the cells by centrifugation at 1500 g for 10 min at 4°C. Resuspend in 1 ml ice-cold TBS and pellet cells as above. Resuspend the cells in TE to a final concentration of 5 x 10^7 cells/ml. Transfer the cells to an Erlenmeyer flask (50 cm^3/1 ml of cells). Add 10 ml lysis solution per ml of cell suspension.③

Notes

① Buffer-saturated phenol should be 'nucleic acid preparation' grade, i.e. colorless. A pink color indicates that it needs to be redistilled.
② For suspension cultures, pellet the cells and wash with TBS.
③ The size of the flask must be increased for large samples to prevent cell aggregation.
④ Do not allow to dry completely.
⑤ Completely degraded DNA runs ahead of the dye.

3 Incubate at 37°C for 1 h. Add 1/200 vol. of proteinase K and mix into the cell suspension using a glass rod. Incubate at 50°C for 3 h with occasional swirling.

4 Cool the solution to room temperature (RT) and phenol extract three times with an equal volume of phenol. Mix the phases by gentle inversion to form an emulsion. Separate the phases by centrifugation at 8000 g for 15 min in a swing-out rotor at RT. Recover the aqueous phase with a wide-bore Pasteur pipette, avoiding the proteinaceous interface.

5 Precipitate the DNA with 0.2 vol. 10 M ammonium acetate and 2.5 vols ethanol and swirl to mix. DNA precipitate will be a visible strand/lump which can be removed to a fresh tube using a thin glass rod or spatula.

6 Wash the DNA precipitate with 70% ethanol and air dry at RT until the ethanol evaporates leaving a moist DNA lump. Redissolve the DNA in 1 ml TE/ 5×10^6 cells and place on an orbital table at 20 r.p.m. for 12–24 h.④

7 Run 1 µl genomic DNA on a 0.5% agarose gel (*Protocol 14*) to assess DNA integrity. Genomic DNA should run as a smear of high molecular weight material.⑤

8 Quantitate 5 µl DNA sample in a 400 µl quartz cuvette at 260 nm according to *Protocol 4*.

Protocol 2. Isolation of genomic DNA from mammalian cells

Protocol 3. Sepharose CL6B spin columns for nucleic acid purification

Reagents

Acid-washed sterile glass beads (No. 11) in $T_{10}E_{0.1}$ (autoclaved)
Autoclaved Sepharose CL6B (1 g/ml) in $T_{10}E_{0.1}$ (10 mM Tris, 0.1 mM ethylenediaminetetra-acetic acid (EDTA); Sigma)

Equipment

Bench top centrifuge
Syringe needles

Technique

1 Pierce the base of a lidless 0.5 ml microcentrifuge tube with a 0.2 mm (bore) syringe needle and add a drop (approx. 20 μl) of glass beads to the base of the tube.①

2 Shake Sepharose CL6B and add to the tube with a Pasteur pipette, filling the tube from the base to the top.②

3 Pierce the base of a lidless 1.5 ml microcentrifuge tube with a 22 gauge syringe needle and place the 0.5 ml microcentrifuge tube into the pierced 1.5 ml microcentrifuge tube.

4 Form the column by centrifugation at 1500 g for 2 min or 14 000 g for 15 sec.③

5 Place the 0.5 ml tube into an unpierced 1.5 ml microcentrifuge tube and add DNA in a total volume of 20–50 μl. Elute desalted DNA by centrifugation at 1500 g for 2 min or 14 000 g for 15 sec.

Notes

① This partially blocks the hole.
② Essential to avoid air bubbles.
③ Bed volume should be 2/3 of height of tube. If less, then Sepharose stock is too dilute. Correct by allowing to settle and discarding excess buffer.

Reagents

Diethyl pyrocarbonate (DEPC)-treated water

Equipment

Quartz cuvettes
UV spectrophotometer

Technique

1 Prior to nucleic acid concentration assessment, 0.5 ml quartz cuvettes should be pre-treated with 0.1% DEPC for 15 min at 37°C to inactivate nucleases. Rinse out the cuvettes with autoclaved water.

2 Turn the spectrophotometer on and adjust the wavelength to 260 nm and leave for 30 min to warm the deuterium lamp. Place 5 µl of the nucleic acid solution into a quartz cuvette and add 395 µl water.

3 Measure absorbance at 260 and 280 nm to assess nucleic acid, polysaccharide and protein concentrations. If the absorbance of sample at 260 nm exceeds 0.8 then dilute the sample until it falls in the range of 0.1 to 0.5 units.

4 Calculate the dilution factor used during the measurement and use this to calculate the nucleic acid concentration where an OD_{260} of 0.5 is equivalent to 50 µg of dsDNA or 40 µg of ssRNA/ssDNA.

III METHODS FOR LABELING NUCLEIC ACIDS

Isotope data (emission strengths, sensitivity, half life)

The half life and emission strength of radioisotopes are important factors to consider during experimental design and during calculation of the effective shelf life of a source isotope. Selected relevant isotopic data is supplied in *Table 2.*

Table 2. Isotopic data

Isotope	Half life	Emission type	Energy of emission (MeV)[1]		Shielding required
			Per particle	Per γ-ray	
^{3}H	12.3 years	β⁻	0.0185	—	None
^{32}P	14.3 days	β⁻	1.71	—	10 mm Perspex
^{33}P	25.4 days	β⁻	0.249	—	None
^{35}S	87.1 days	β⁻	0.169	—	None
^{125}I	60 days	γ	—	0.035	Lead-impregnated acrylic sheet

Choice of labeling method

Probes can be radiolabeled or labeled non-isotopically using biotinylation. The probe may be either single-stranded DNA or RNA. The advantages of an RNA probe is that RNA:RNA hybrids are more stable than DNA:RNA and DNA:DNA hybrids. The choice of probe used depends on: (i) target abundance, (ii) availability of information on target species, (iii) sensitivity required, (iv) requirements for multiple probings of DNA/RNA sequences, and (v) safety aspects in probe production. These factors are discussed below.

Target abundance

The abundance of the target species under study to a large extent defines which probes can be used in hybridization studies. Highly abundant species can be studied using less sensitive techniques as well as highly sensitive methods (if desirable). *Table 3* details the typical minimum amounts of DNA or RNA detectable on blots by each of the three most popular methodologies. Rarer RNA species must be studied with more sensitive techniques in order to detect the formation of hybrids.

Table 3. Sensitivity of selected probing systems

System basis	Sensitivity (ng/mm^2 membrane)	Isotopic/nonisotopic
Biotin-avidin-HRP[a]	0.5	Nonisotopic
^{32}P random primed probes	0.1	Isotopic
^{32}P riboprobes	0.05	Isotopic

[a]HRP-Horseradish peroxidase.

Availability of information on target species

Information on the target species sequence determines the range of probes that can be used. If complete sequence identity of the RNA species is known and complementary DNA templates are available then there is no restriction on the probe type that can be used. For example, if a certain mRNA species is under study and the cDNA is available, then preferred methods would include random priming, radiolabeled RNA probe generated from a transcription site in the vector containing the cDNA or a biotinylated DNA fragment/oligonucleotide.

Sensitivity required

Requirements of experiments can include detection and/or quantitation of one or a series of DNA or RNA species. If detection of a single DNA or RNA species is the only requirement then the sensitivity of the probe used is only defined by the limit imposed by DNA or RNA target abundance. For example, detection of a highly abundant RNA species in a sample can be accomplished by any of the procedures listed in *Table 3*. However, a low sensitivity method would be favored in order to reduce the manipulations involved in the hybridization. A more sensitive method is required to detect a DNA or RNA species that is in low abundance. If detection is not the only requirement of the experiment, but quantitation is the emphasis, then a highly sensitive method is required to ensure that the signal produced at the end of the hybridization procedure can be related directly to the abundance of DNA or RNA sequence.

Requirements for multiple probings of DNA or RNA sequences

Hybridization involving multiple probings of DNA or RNA on membranes is usually accomplished by using DNA probes, and it is important to use probes that can easily be removed, allowing further probings without leaving a confusing background of the old signal. Radioactive probes (e.g. random primed synthesis) or nonisotopic probes can both be used in multiple probing procedures. It may be noted here that if the sample of DNA or RNA is plentiful then multiple probing of the same samples is often not required as aliquots may be probed in isolation with different probes. This is necessary when carrying out multiple probings of RNA in solution using RNA probes (riboprobes) which require multiple samples as the method digests RNA that is not bound to the probe. Riboprobe methods therefore require a relatively large amount of RNA for multiple probings.

Probes can be removed from a membrane after autoradiography by 'stripping' the label from the membrane. Thus probes that are easier to strip from a membrane (small probes and those that hybridize to low copy number targets) should be the initial probes and probes producing large signal responses should be used at the end of the series of hybridizations. Nonisotopic probings have the advantage of not having to be stripped from the membrane. Chemiluminescent signals (e.g. derived from the dioxethane ion destabilization or luminol hydrolysis) are short lived (maximum of 1 h) and further probings can be carried out as soon as the label has been exhausted.

Safety aspects in probe production

Because of the inherent risks in using radioisotopes, the use of radioactive materials is regulated in most countries (see also Safety section on p. xii). It is important that the reader is aware of the controls on the use of radioisotopes and does not contravene regulations. The risks of working with radioactivity can be minimized by keeping personal exposure times as short as possible and the amount of isotope as low as possible. When working with isotopes such as ^{32}P and ^{125}I it is essential to use shielding; Perspex for the former and lead for the latter. Given the risks associated with using radioisotopes, it is not surprising that many laboratories are now using nonisotopic methods for hybridization studies.

The following guidelines should be used during isotope handling procedures.

(i)   Consult the appropriate radioisotopes supervisor to familiarize yourself with all procedures relevant to your research institute.

(ii)   Always defrost all radioactive solutions/sources for at least 20 min prior to use to avoid radioactive aerosols which are produced when opening a frozen source.

(iii) Mix source very gently but thoroughly using a pipette tip prior to use.

(iv) When boiling isotopes (heat denaturing) use screw-capped microcentrifuge tubes to prevent the lid bursting open under pressure and producing aerosols.

(v)   If a bag system is being used during hybridization protocols, place the sealed bag containing the isotope inside of another bag and seal it to reduce the risk of spillage.

The *amount* of probe in a purified DNA or RNA sample, the incorporation *efficiency* and its *specific activity* can then be calculated as follows:

Quantification of the *amount* of a nucleotide label in a probe after Cerenkov counting:
- Cerenkov count a fraction of the probe of a ^{32}P-labeled DNA/RNA sample.
- If a sample of 5 µl has a count of 100 000 c.p.m. then adjusting for 50% counting efficiency gives an activity of 200 000 c.p.m.
- If total volume of the isolated and purified probe was 50 µl then the total activity was 2×10^6 c.p.m.

The *efficiency* of incorporation for this probe can then be calculated:

- If the initial probe production protocol used 5 µl of $[\alpha\text{-}^{32}P]$UTP equivalent to 1.85 MBq/50 µCi (1.1×10^8 d.p.m.) then the efficiency of incorporation was $\dfrac{2 \times 10^6 \text{ d.p.m.} \times 100}{1.1 \times 10^8 \text{ d.p.m.}} = 1.8\%$.

The *specific activity* of the probe:

- The concentrations of nonradioactive UTP and radioactive $[\alpha\text{-}^{32}P]$UTP need to be known.
- For example, in the preparation of a riboprobe 2.5 µl of a 0.1 mM UTP (2.5×10^{-10} mol) and 5 µl of 25 µM $[\alpha\text{-}^{32}P]$UTP (1.25×10^{-10} mol) were used then the total amount of UTP in the reaction was 3.75×10^{-10} mol.
- If 1.8% of the 3.75×10^{-10} mol of UTP was incorporated then 6.8×10^{-12} mol of UTP was present in the probe.
- If the probe length was 1000 nt and uracil occurs once every four bases on average (250 nt), then the molar amount of RNA transcript synthesized is: $\dfrac{6.8 \times 10^{-12} \text{ mol}}{250} = 2.7 \times 10^{-14}$ mol = 27 fmol.
- The 1000 nt transcript has a molecular weight of $330 \times 1000 = 330\,000$, 1 mol of 1000 nt = $330\,000$ g and 27 fmol = $330\,000 \times 2.7 \times 10^{-14} = 8.9 \times 10^{-3}$ µg. The specific activity of the RNA probe is then $\dfrac{2 \times 10^6 \text{ d.p.m.}}{8.9 \times 10^{-3} \mu g} = 2.2 \times 10^8$ d.p.m./µg.

Types of nonradioisotopic labeling

Chemiluminescence is becoming a favored method of localization and assessment of nucleic acids. The procedure centers on conjugation of biotin to the protein avidin. Avidin can then be used as an epitope site for anti-avidin antibodies. These antibodies can either have a reporter enzyme attached, such as horseradish peroxidase (HRP) or alkaline phosphatase, or can form the epitope for a secondary antibody which has a reporter enzyme present. Biotin can be conjugated to nucleic acids by a method called biotinylation. This may be attached to nucleic acids enzymatically by incorporation of biotin-tagged nucleotides or chemically to preformed oligonucleotides by photobiotinylation. Hybridization of biotinylated nucleic acid specific to the RNA of interest on a northern blot containing the RNA allows localization of RNA bands by staining or chemiluminesence.

A commercial system which uses HRP as a reporter enzyme is called enhanced chemiluminescence (ECL) and can be purchased from Amersham International. This system, in the presence of hydrogen peroxide, can oxidize the substrate luminol. Luminol is raised to an excited state by this reaction and returns to its ground state by the emission of light. The enhancement integral to the system is accomplished by a phenolic enhancer which increases the intensity of emitted light by up to 1000 times. The reaction peaks after 1–5 min and then declines. A permanent record can be made by placing a piece of X-ray film over the membrane for a few seconds (exposure time varies with emission intensity) and developing immediately. Various exposures are performed at the one time to ensure that a correct exposure is achieved.

The other main reporter gene, alkaline phosphatase, can be used to cleave substrates with a concomitant emission of light. The two substrates commonly used are disodium 3-(4-methoxyspiro[1,2-dioxethane-3,2′-tricyclo-[3.3.1.1^{3,7}]decan]4-yl)phenyl phosphate (AMPPD) and a 5-chloro derivative (CSPD) (Beckman). The reporter enzyme dephosphorylates the substrate, destabilizing it. Light is emitted as the molecule breaks down. Enhancers for the emission are also available for this system. A commercial system, Quicklight™, using this reporter gene is available from FMC (see Appendix C).

Methods available

Radiolabeling

Refer to safety procedures section prior to any manipulation of radioisotopes.

3′-End labeling of DNA (see *Protocol 5*)

This procedure is usually carried out using the enzyme terminal deoxynucleotidyl kinase (terminal transferase) which catalyzes the

References

1. Cozzarelli, N.R., Kelly, R.B. and Kornberg A. (1969) *J. Mol. Biol.* **45**:513.
2. Tu, C.-P.D. and Cohen, S.N. (1980) *Gene*, **10**:177.
3. Deng, G. and Wu, R. (1983) *Methods Enzymol.* **100**:96.
4. Michelson, A.M. and Orkin, S.H. (1982) *J. Biological Chemistry*, **257**:14773.
5. Nelson, T. and Brutlag, D. (1979) *Methods Enzymol.* **68**:41.

repetitive addition of mononucleotides using dNTP substrates to the 3′-terminus of a DNA primer. This addition can produce a labeled tail [1] or a single incorporated nucleotide when chain terminators are used (ddNTPs) [2].

Problems: *Protocol 5* may also be applied to double-stranded DNA, preferably with a protruding 3′-terminus. DNA fragments possessing recessed 3′-termini or blunt-ended molecules will not be uniformly labeled using this method and are better labeled by the procedures detailed in refs 3–7.

3′-End labeling of RNA (see *Protocol 6*)

This is usually carried out using the enzyme RNA ligase but can be accomplished using terminal deoxynucleotidyl kinase (terminal transferase) or poly(A) polymerase (detailed in ref. 2 of Chapter II).

Problems: RNA degradation must be avoided throughout the procedure.

5′-End labeling of DNA (see *Protocol 7*)

Polynucleotide kinase is used to 5′-end label DNA according to *Protocol 7* [8 and 9] prior to procedures such as Maxam–Gilbert sequencing [10], nuclease SI analysis (*Protocol 18*), or northern analysis using oligonucleotide probes.

Problems: Dephosphorylated DNA should be gel purified to remove low molecular weight nucleic acids that would otherwise be preferentially labeled. If DNA is precipitated prior to the procedure then it

6. Roychoudhury, R., Jay, E. and Wu, R. (1976) *Nucleic Acids Res.* **3**:101.

7. Roychoudhury, R. and Wu, R. (1980) *Methods Enzymol.* **65**:43.

8. Richardson, C.C. (1971) In *Procedures in Nucleic Acid Research* (Cantoni, G.L. and Davies, D.R., eds) vol **2**:815. Harper and Row, New York.

9. Berkner, K.L. and Folk, W.R. (1977) *J. Biological Chemistry*, **252**:3176.

10. Maxam, A.M. and Gilbert, W. (1977) *Proc. Natl Acad. Sci., USA*, **74**:560.

11. Furuichi, Y. and Shatkin, A.J. (1994) In *RNA Processing – A Practical Approach,* volume II (Higgins, S.J. and Hames, B.D., eds.) p.46. IRL Press at Oxford University Press, Oxford.

12. Unhlenbeck, O.C. and Gumport, R.I. (1982) In *The Enzymes*, 3rd edition (P.D. Boyer, ed.), vol. 14, p. 87. Academic Press, New York.

13. Shinshi, H., Miwa, M., Sugimura, T., Shimotohno, K. and Miura, K. (1976) *FEBS Lett.* **65**: 251.

14. Maniatis, T., Jeffrey, A. and Kleid, D.G. (1975) *Proc. Natl. Acad. Sci., USA*, **72**:1184.

15. Rigby, P.W.J., Dieckmann, M., Rhodes, C. and Berg, P. (1977) *J. Mol. Biol.* **113**:237.

16. Lamond, A.I. and Sproat, B.S. (1994) In *RNA Processing – A Practical Approach* (Higgins, S. J. and Hames, B. D., eds). IRL Press at Oxford University Press, Oxford.

should be precipitated with sodium acetate not ammonium acetate as ammonium ions inhibit T4 polynucleotide kinase.

5′-End labeling of RNA (see Protocols 8a and b)

The 5′-terminus of most RNA molecules is phosphorylated; however, eukaryotic mRNA (except mitochondrial mRNA) usually has a cap structure which must be considered when labeling its 5′-terminus. The 5′-termini of other RNA types may be directly labeled (start at *Protocol 8b*). *Protocol 8a* uses the enzyme tobacco acid pyrophosphatase to decap the mRNA which can also be accomplished chemically according to ref. 11.

Problems: RNA degradation must be avoided throughout the procedure.

Nick translation (see Protocol 9)

The enzyme DNA polymerase I is used in this method to add nucleotides to free 3′-hydroxyl termini that are available after DNA has been nicked. The enzyme displaces and removes nucleotides 5′ to the incorporation site and this displacement and incorporation is called nick translation. Using this method it is possible to produce DNA labeled to a specific activity of 10^8 c.p.m./μg. This method yields probes that are not as large as those prepared by *in vitro* transcription or PCR (*Protocols 10 and 12*). Larger probes are necessary for some procedures, for example analysis of DNA:RNA hybrids using nuclease SI.

17. Feinberg, A.P. and Vogelstein, B. (1983) *Analytical Biochemistry,* **132:** 6
18. Garbarino, J.E., Rockhold, D.R. and Belknap, W.R. (1992) *Plant Mol. Biol.* **20:**235-244.
19. ECL direct nucleic acid labeling and detection systems booklet. Amersham Life Science.

Protocols provided

Methods for labeling nucleic acids

Problems: DNA preparations containing impurities, particularly from agarose gel electrophoresis elutions are inefficiently labeled. Therefore ensure DNA is purified prior to labeling (see *Protocol 3*). The probes produced by this method are of high specific activity which can lead to radioactive decay damaging the probes. The probes should therefore be used soon after preparation.

Riboprobe preparation by **in vitro** *transcription* (see *Protocol 10*)
Riboprobes are usually produced from plasmid templates containing transcription start sites for SP6 or T7 bacteriophage DNA-dependent RNA polymerase. This technique yields probes of high specific activity and is quite easy to carry out.
Problems: The probes produced by this method are of high specific activity which can lead to radioactive decay damaging the probes. The probes should therefore be used soon after preparation.

Random hexamer primed synthesis (see *Protocol 11*)
Generation of probes using the random hexamer primed synthesis method has been the favored protocol in laboratories during the past decade but may soon be totally replaced by PCR-based methods. The method relies on random primers (usually hexamers) binding to DNA templates allowing primed synthesis to copy the DNA template, incorporating radionucleotides in the process.
Problems: Templates smaller than 200 bp are inefficiently labeled and

the probe must be purified by column chromatography (see *Protocol 3)* prior to use in hybridization studies.

PCR production of [³²P]CTP-labeled fragments (see *Protocol 12*)

This method produces probes of high specific activity and purity from 30–2000 bp. This method may become the methodology of choice over the next few years as the preparation is easy.

Problems: The probes produced by this method are of high specific activity which can lead to radioactive decay damaging the probes. The probes should therefore be used soon after preparation.

Nonradioisotopic labeling

Enhanced chemiluminescence (ECL) (see *Protocol 13*)

This method of labeling uses an enzyme-labeled probe to produce light upon catalysis of a chemical substrate. The light emission is recorded on X-ray film in much the same way as radioisotopic emissions.

Protocol 5. Labeling 3′-termini of single-stranded DNA primers using terminal transferase

Reagents

[α-^{32}P]dATP (29.6 TBq (800 Ci)/mmol, 370 MBq (10 mCi)/ml)△
500 µg/ml Bovine serum albumin (BSA)
Sterile double-distilled water
Terminal transferase (15–30 U/µl; Promega)
5× Terminal transferase buffer (500 mM cacodylate, pH 6.8,
  1 mM CoCl$_2$, 0.5 mM dithiothreitol (DTT))

Equipment

14 000 g microcentrifuge
Perspex shielding
Sepharose CL6B column (see *Protocol 3*)
Vortex mixer
Water baths at 37, 70 and 100°C

Technique (ref. 1)

1 Set up the labeling reaction:

- 4 µl 5× terminal transferase buffer
- 10 pmol 3′OH ends (Primers; see p. 149)
- 1.6 µl [α-^{32}P]dATP
- 1 µl terminal transferase

and adjust the final volume to 20 µl using sterile double-distilled water.①

2 Incubate at 37°C for 30–60 min and terminate labeling reaction by heating to 70°C for 10 min.②

Notes

① **Caution:** Defrost and mix [α-^{32}P] dATP prior to use.
② **Caution:** Heating samples to 100°C can cause a rise in pressure in the microcentrifuge tube containing the sample. A sealed screw threaded tube should be used which will not pop open.

3 Add 100 µl sterile double-distilled water and heat the probe to 100°C
   for 2 min. Cool on ice and collect any condensation by centrifugation.
   Purify the probe by passage through a Sepharose CL6B spin column
   (*Protocol 3*) of Sephadex G-50 spin columns (Pharmacia) prior to heat
   denaturation and addition to hybridization solution.

Protocol 6. 3′-End labeling of RNA using T4 RNA ligase

Reagents

0.25 mM Adenosine triphosphate (ATP)
5 M Ammonium acetate (pH 7.0)
0.1 mg/ml Bovine serum albumin (BSA)
Dimethylsulfoxide (DMSO)⚠
Ethanol⚠
70% Ethanol⚠
Glycogen (20 mg/ml)
10× Ligase buffer

[5′-^{32}P]pCp (111 TBq (3000 Ci)/mmol at 370 MBq (10 mCi)/ml)⚠
Sterile water
T4 RNA ligase (50 units/µl)

Equipment

14 000 g microcentrifuge
Perspex shielding
Vortex mixer
Water bath at 37°C/56°C

Technique

1 Add 0.1 vol. 5 M ammonium acetate (pH 7.0), 0.1 vol. glycogen and 3 vols ethanol to precipitate the RNA. Place at –20°C for 3 h before collecting by centrifugation at 14 000 g for 10 min at RT.①☐1

2 Wash the RNA pellet with 10 µl 70% ethanol and air dry for 2 min. Redissolve RNA in water to a final concentration of 5 pmol/µl.②

3 Place a microcentrifuge tube on ice and add 10 pmol RNA substrate (e.g. 2 µl 5 pmol/µl), 2 µl 10× ligase buffer, 2 µl 0.1 mg/ml BSA, 2 µl DMSO, 2 µl 0.25 mM ATP, 2 µl [5′-^{32}P]pCp and 0.5 µl (20 units) T4 RNA ligase. Add sterile water to a final volume of 20 µl.③

Notes

① Steps 1–2 may be skipped if RNA concentration >5 pmol/µl.

② **Caution:** Defrost and mix [5′-^{32}P]pCp prior to use.

③ The probe produced should have a specific activity of >10^6 c.p.m./pmol. This will be lower if there is a uridine monophosphate (UMP) at the 3′ terminus.

④ **Danger:** Electrophoresis of radioactive probes is hazardous – extreme care needed.

4 Place in a 4°C fridge for 10–12 h. Isolate the labeled RNA by denaturing polyacrylamide gel electrophoresis and RNA elution (*Protocols 16* and *33*) with adequate shielding.④

Pause point

1 May be left at –20°C for up to 24 h.

Protocol 7. 5′-End labeling of DNA – phosphorylation method

Reagents

7.5 M Ammonium acetate
[γ-^{32}P]ATP (111 TBq (3000 Ci)/mmol at 370 mBq (10 Ci)/ml)⚠
Calf alkaline intestinal phosphatase (CAIP)
Chloroform–isoamylalcohol (IAA) (24:1)⚠
Dephosphorylation buffer (500 mM Tris-HCl (pH 9.0), 10 mM MgCl$_2$, 1 mM ZnCl$_2$, 10 mM Spermidine)
0.5 M Ethylenediaminetetra-acetic acid (EDTA)
Ethanol⚠
Forward exchange buffer (500 mM Tris-HCl, 100 mM MgCl$_2$, 50 mM dithiothreitol (DTT), 1 mM spermidine)

2 M NaCl
Phenol–chloroform–IAA (25:24:1)⚠
TE (10 mM Tris, 1 mM EDTA, pH 7.5)
T4 polynucleotide kinase (8–10 units/µl)

Equipment

14 000 g microcentrifuge
Perspex shielding
Sepharose CL6B spin column (*Protocol 3*)
Vortex mixer
Water bath at 37°C/56°C

Technique (refs 8 and 9)

1 Set up the dephosphorylation reaction by mixing:
- 5 µl dephosphorylation buffer
- 1 µl purified substrate DNA (≤20 pmol of 5′ ends; see p. 149)①
- 0.5 µl CAIP②
- 43.5 µl water

2 *Protruding 5′-termini*: Incubate at 37°C for 30 min before adding a further 0.1 units of CAIP and incubate for a further 30 min at 37°C.

Notes

① Purify DNA prior to procedure according to *Protocol 3*.
② Routinely test CAIP enzyme on a nonessential template to ensure the absence of exonuclease activity.
③ **Caution:** Defrost and mix [γ-^{32}P]ATP prior to use. ATP concentration used should be at least 1 µM.

Recessed 5′-termini: Incubate at 37°C for 15 min before adding a further 0.1 units of CAIP and incubating at 37°C for a further 15 min.

Blunt-ended molecules: Incubate at 56°C for 15 min before adding a further 0.1 units of CAIP and incubating at 56°C for 15 min.

3 Terminate the reaction by extracting the proteins from the reaction mixture with 50 μl of phenol–chloroform–IAA (25:24:1). Vortex for 1 min and centrifuge at 12 000 g for 2 min.

4 Recover the upper aqueous layer into a fresh microcentrifuge tube. Extract the trace levels of phenol by addition of 50 μl chloroform–IAA (24:1) extraction. Vortex for 30 sec and centrifuge at 12 000 g for 2 min.

5 Precipitate the DNA from the recovered aqueous fraction by addition of 0.1 vol. 2 M NaCl and 2 vols ethanol and place at –20°C for 30 min. Pellet the DNA by centrifugation at 12 000 g for 5 min and discard the supernatant. Resuspend the DNA in 34 μl forward exchange buffer.

6 Set up the phosphorylation reaction by mixing:
   - 15 μl [γ-^{32}P]ATP③
   - 1 μl T4 polynucleotide kinase
   - 1 μl dephosphorylated DNA (=5 pmol 5′ ends; see p. 149).

7 Incubate at 37°C for 10 min and terminate the reaction by addition of 2 μl 0.5 M EDTA. Extract the proteins from the reaction mixture by addition of an equal volume of phenol chloroform–IAA (25:24:1). Vortex for 1 min and centrifuge for 2 min at 12 000 g at RT.

Continued overleaf

Protocol 7. 5′-End labeling of DNA

8 Precipitate the DNA in the recovered aqueous fraction with 0.5 vol. 7.5 M ammonium acetate and 2 vols ethanol and place at −20°C for 30 min.1

9 Pellet the DNA by centrifugation at 12 000 *g* for 5 min and redissolve in 50 µl TE. Probes are purified by passage through Sepharose CL6B spin columns (*Protocol 3*) or Sephadex G-50 columns.

Pause point

1 May be left at −20°C for up to 24 h.

Protocol 8a. Decapping of mRNA using tobacco acid pyrophosphatase

Reagents

Calf alkaline intestinal phosphatase (CAIP)
100 mM Dithiothreitol (DTT)
0.5 M Sodium acetate (pH 6.0)
0.25 M Potassium phosphate (pH 9.5)
Tobacco acid pyrophosphatase (2.9 units/μl)
0.5 M Tris-HCl (pH 8.0)

Equipment

Boiling water bath
Microcentrifuge
Water baths at 37°C and 50°C

Technique (ref. 13)①

1 To 100 nmol RNA in a volume of 7 μl add 1 μl 0.5 M sodium acetate (pH 6.0).

2 Heat denature the RNA at 100°C for 2 min and snap cool on ice and collect the condensation by centrifugation for a few seconds.

3 Decap the mRNA by adding 1 μl 100 mM dithiothreitol (DTT) and 1 μl (2.9 units) of tobacco acid pyrophosphatase. Incubate at 37°C for 30 min.

4 Dephosphorylate the decapped mRNA by adding 2 μl 0.5 M Tris-HCl (pH 8.0) and 1 μl (20 mU) CAIP (Boehringer Mannheim), incubating at 50°C for 30 min.

5 Stop the reaction by addition of 1 μl 0.25 M potassium phosphate (pH 9.5).

Notes

① This procedure relies on removal of the cap nucleotide by tobacco acid pyrophosphatase with subsequent dephosphorylation by alkaline phosphatase.

Protocol 8b. Phosphorylation of RNA using polynucleotide kinase

Reagents

4 M Ammonium acetate (pH 7.0)
$[\gamma\text{-}^{32}P]$ATP (111 TBq (3000 Ci)/mmol at 370 MBq (10 mCi)/ml)⚠
100 mM Dithiothreitol (DTT)
Ethanol⚠
0.2 M $MgCl_2$, 32 mM spermidine
Polynucleotide kinase (4 units/µl)

Equipment

Boiling water bath
Microcentrifuge
Perspex shielding
Sepharose CL6B spin column (*Protocol 3*)
Water baths at 37°C at 50°C

Technique①

1. Heat denature the RNA at 100°C for 2 min and snap cool on ice. Collect RNA at the base of the tube by brief centrifugation.

2. Set the labeling reaction up:
   - 1 µl 100 mM DTT
   - 1 µl of 0.2 M $MgCl_2$, 32 mM Spermidine (Sigma)
   - 1 µl of $[\gamma\text{-}^{32}P]$ATP②
   - 1 µl (4 units) of polynucleotide kinase and incubate at 37°C for 30 min.

3. Stop the reaction by adding 16 µl 4 M ammonium acetate (pH 7.0) and 80 µl of ethanol and place at −20°C overnight, precipitating the RNA.

4. Collect the precipitate by centrifugation at 14 000 *g* at −4°C for 20 min. Discard the supernatant containing unincorporated $[\gamma\text{-}^{32}P]$ATP. Purify probes by passage through Sepharose (*Protocol 3*) or Sephadex columns.

Notes

① This phosphorylation protocol may be applied to any dephosphorylated RNA.

② **Caution:** Defrost $[\gamma\text{-}^{32}P]$dATP for at least 20 min to avoid radioactive aerosols upon opening and then mix radioisotope prior to use by gentle stirring with a pipette tip.

Protocol 9. Radiolabeling of DNA using nick translation

Reagents

$[\alpha\text{-}^{32}P]dCTP$ (14.8 TBq (400 Ci)/mmol at 370 MBq (10 mCi)/ml)⚠
10× Nick translation buffer (500 mM Tris-HCl (pH 7.2), 100 mM MgSO$_4$, 1 mM dithiothreitol (DTT))
Nucleotide mix (500 μM each dNTP omitting radiolabeled one)
Optimized enzyme mix (DNase I, DNA polymerase; Promega)
Stop mix (0.2 M ethylenediaminetetra-acetic acid (EDTA), pH 8.0)

Equipment

Boiling water bath
Incubator at 15°C
Microcentrifuge
Perspex shielding
Sepharose CL6B spin columns (*Protocol 3*)

Technique (refs 14 and 15)

1 Set up the labeling reaction by mixing
   - 1 μg template DNA in a maximum volume of 23 μl
   - 5 μl 10× nick translation buffer
   - 10 μl nucleotide mix
   - 7 μl $[\alpha\text{-}^{32}P]dCTP$①
   - 5 μl optimized enzyme mix
   Make up the reaction volume to 50 μl with sterile, double-distilled water.

2 Incubate at 15°C for 1 h to allow nicking of DNA and repair by DNA polymerase, thus incorporating radiolabeled nucleotides.

3 Terminate the reaction by the addition of 5 μl of stop mix (0.2 M EDTA, pH 8.0).

Notes

① **Caution:** Defrost $[\alpha\text{-}^{32}P]dCTP$ for at least 20 min to avoid radioactive aerosols upon opening. Mix defrosted radioactivity prior to removing the required 7 μl by gentle stirring with a pipette tip.

② See *Protocol 3* for Sepharose spin column preparation.

③ **Caution:** A sealed screw-threaded tube should be used which will not pop open.

4 Remove the unincorporated radiolabel by gel filtration using a Sepharose CL6B column or Sephadex G-50 (Pharmacia) column.②

5 Denature the DNA by incubating in a microcentrifuge tube at 100°C for 5 min and then transfer the tube to ice (preventing re-annealing).③
This method usually incorporates more than 65% of the label into the DNA.

Protocol 10. Riboprobe preparation by *in vitro* transcription

Reagents

[α-^{32}P]rUTP (specific activity of 29.6 TBq (800 Ci)/mmol, 20 mCi/ml)⚠

Chloroform–isoamyl alcohol (IAA) (24:1)⚠

0.1 M Dithiothreitol (DTT)

DNase (RNase-free DNase; Boehringer Mannheim)

Phenol–chloroform–IAA (25:24:1)⚠

Ribonucleotide mix (each of rATP, rCTP and rGTP at 2mM and 0.1 mM rUTP)

Rnasin (1unit/μl; Promega)

SP6, T3 or T7 bacteriophage DNA-dependent RNA polymerase

10× Transcription buffer (40 mM Tris-acetate (pH 7.5), 0.5 M sodium acetate (pH 7), 80 mM magnesium acetate, 20 mM Spermidine)

Vector DNA containing insert

Equipment

Boiling water bath

Microcentrifuge

Perspex shielding

Sepharose CL6B spin columns (*Protocol 3*)

Water baths at 37°C and 40°C

Technique (ref. 16)

1 Linearize 0.1 μg plasmid DNA using restriction sites in the polylinker flanking the insert to be transcribed.①

2 Extract with equal volumes of phenol–chloroform–IAA (25:24:1) to remove the enzymes used in linearization and blunting or PCR procedures. Vortex for 30 sec, centrifuge at 14 000 *g* for 3 min and recover the supernatant.

3 Chloroform extract with an equal volume of chloroform–IAA (24:1),

Notes

① Linearized DNA may be generated by restriction digestion of vectors or by a PCR reaction.

② **Caution:** Defrost and mix [α-^{32}P]rUTP before use otherwise the solution is not homogeneous and opening a frozen bottle can give rise to radioactive aerosols.

vortex for 30 sec, centrifuge at 14 000 g for 3 min and recover the supernatant.

4 Set up the transcription reaction:
   - 10 µl 10× transcription buffer
   - 10 µl of ribonucleotide mix
   - 0.5 µg linearized plasmid DNA
   - 6 µl [α-^{32}P]rUTP②
   - 10 µl Rnasin
   - 10 µl 0.1 M DTT
   - 50 units bacteriophage DNA-dependent RNA polymerase
   to a final volume of 100 µl.

5 Transcription proceeds for 1–2 h at 37°C (T3 and T7 DNA-dependent RNA polymerases) or 40°C (SP6 DNA-dependent RNA polymerase).

6 Add 2.5 µl DNase (10 units/µl) and incubate at 37°C for 30 min.

7 Phenol-extract the sample and pass through a CL6B (*Protocol 3*)/ Sephadex G-50 spin column prior to hybridization studies.

Protocol 10. Riboprobe preparation by in vitro *transcription*

Protocol 11. **Random hexamer primed synthesis**

Reagents

[α-^{32}P]dATP (222 TBq (6000 Ci)/mmol; 740 MBq (20 mCi)/ml)△
6 units/µl Klenow (large fragment of DNA polymerase)
'Master-mix' (43.8 µM dCTP, dGTP, dTTP, 0.438 M Hepes, 12.3 units/ml random hexanucleotide; Pharmacia)
Stop mix (50 mM EDTA, 1 mg/ml calf thymus DNA, 1 mg/ml dextran 804 (Sigma), 0.1% (w/v) bromophenol blue)

Equipment

Boiling water bath
Ice
Microcentrifuge
Perspex shielding
Sepharose CL6B spin columns (*Protocol 3*)

Technique (ref. 17)

1 Heat denature 25–100 ng of the dsDNA fragment in a volume of 11 µl at 100°C for 5 min. Prevent the separate strands from re-annealing by transferring the sample tube into ice for 3 min ('snap cooling').①

2 Add an equal volume (11 µl) of 'master-mix', 0.5 µl Klenow (3 units) and 2 µl [α-^{32}P]dATP. Leave at RT to allow synthesis of the new DNA fragments for 2–5 h.②

3 Terminate synthesis by adding 5 µl stop mix.

4 Purify the labeled products from free nucleotides by passage through a Sepharose CL6B spin column (*Protocol 3*).

5 Add 100 µl sterile double-distilled water to the purified DNA and heat

Notes

① Template DNA should be at least 200 nt in length, as smaller templates lead to the production of short probes and a resultant high background.

② **Caution:** [α-^{32}P]dATP is a radioactive source emitting at a high intensity. Shielding at all times is imperative. Primary source nucleotide should be removed from the freezer at least 30 min prior to use to allow thawing. Opening frozen primary sources can create radioactive aerosols.

denature the labeled strands from the template DNA at 100°C for 5 min. Place on ice to snap cool, preventing re-annealing.

6 The probe may then be introduced with hybridization solution to the membrane.

Protocol 12. PCR production of [³²P]CTP-labeled fragments

Reagents

$[\alpha\text{-}^{32}P]dCTP$ (14.8 TBq (400 Ci)/mmol, 370 MBq (10 mCi)/ml)⚠
Chloroform–isoamyl alcohol (IAA) (24:1)⚠
10 mM dATP, dGTP, dCTP and dTTP
Mineral oil
20 μM stocks of primer 1 and 2 (e.g. T3 and T7 primers)
10× Taq polymerase reaction buffer (Pharmacia)

5 units/μl Taq polymerase (Pharmacia)
Template DNA (3 ng)

Equipment

Scintillation counter
Sepharose CL6B spin column (*Protocol 3*)
Shielding
Thermocycler

Technique (ref. 18)

1 In a 0.5 ml microcentrifuge tube, mix the following:
   - 10.0 μl 10× reaction buffer
   - 2 μl 10 mM dATP (final conc. 200 μM)
   - 2 μl 10 mM dGTP (final conc. 200 μM)
   - 2 μl 10 mM dTTP (final conc. 200 μM)
   - 2 μl 25 μM dCTP (for fragment <150 bp)①
   - 5 μl (1 MBq/50 μCi) $[\alpha\text{-}^{32}P]dCTP$②
   - 0.5 μl (2.5 units) Taq polymerase
   - 3 μl primer 1 (20 μM) (final conc. 0.6 μM)
   - 3 μl primer 2 (20 μM) (final conc. 0.6 μM)
   - 3 μl (3 ng) template

Notes

① For 17-mer, 20 μM = 110 μg/ml.
   Optimum incorporation:
   130 bp 2 μl 25 μM cold dCTP; 700 bp 2 μl 0.3 mM cold dCTP; 980 bp 2 μl 0.3 mM cold dCTP. Optimize to your system by using diluted 10 mM stock according to fragment size.

② **Caution:** $[\alpha\text{-}^{32}P]dATP$ is a radioactive source emitting at a high intensity. Shielding at all times is imperative. Primary source nucleotides should be removed from the freezer at least 30 min prior to use to allow thawing. Opening frozen primary sources can create radioactive aerosols.

- 67.5 µl sterile double-distilled water
  and add 100 µl of mineral oil on top of reaction.

2 Carry out PCR using the following program:
  - 94°C, 1 min pre-heat
  - 94°C, 1 min denaturation
  - 45–50°C, 1 min anneal/25 cycles
  - 72°C, 0.5 to 1 min extension
  - 72°C final extension for 2 min

3 When PCR is finished, add 100 µl chloroform–IAA and 25 µl water to the reaction tube and vortex. Recover the aqueous (upper) layer and pass through a spin column (*Protocol 3*).

4 Count 1 µl product in a scintillation counter. There should be about $1–1.5 \times 10^5$ c.p.m./µl.

5 To check the reaction products, run 1 µl on small vertical gel of 3% acrylamide gel/1% agarose gel (*Protocol 16*). Cover the wet gel with plastic wrap and expose to film for about 10 min.

Reagents

Supplied in ECL kit:
Labeling reagent
Glutaraldehyde solution⚠
To be supplied by user:
Sterile glycerol

Equipment

Boiling water bath
Ice
Sepharose CL6B spin columns (*Protocol 3*)
Water bath at 37°C

Technique (ref. 19)

1 Desalt the probe DNA/RNA stock by passage through Sepharose CL6B spin columns (*Protocol 3*). Dilute the probe DNA to 10 ng/μl with water.

2 Denature 100 ng (10 μl) of the DNA sample in a screw top microcentrifuge tube by heating for 5 min in a boiling water bath.①

3 Snap cool the sample by placing immediately on ice for 5 min and collect by brief centrifugation. Add 10 μl labeling reagent and mix.

4 Add 10 μl glutaraldehyde solution and mix. Collect at the base of the tube by brief centrifugation. Incubate in a 37°C water bath for 10 min.

5 If the probe is not to be used immediately then place it on ice for up to 15 min.

6 The probe can be stored for up to six months by the addition of 30 μl glycerol and mixing. This 50% glycerol solution may be kept at −15°C to −30°C.②

Notes

① Screw top microcentrifuge tubes will not pop open when boiled.

② Probes of less than 300 bases should be incubated for 20 min.

Protocol 13b. **ECL hybridization and washing**

Reagents

Supplied in ECL kit (Amersham International):
Blocking agent
Hybridization buffer
To be supplied by user:
NaCl (analytical grade)
Primary wash buffer (0.4% SDS, 0.5× SSC (with 6 M urea for urea wash))
Secondary wash buffer (20× SSC (3 M NaCl, 300 mM tri-sodium citrate, pH adjusted to 7.0))

Equipment

Hotplate magnetic stirrer (42°C)
Hybridization vessels
Orbital incubators (42/55°C)
Water bath (42/55°C)

Technique

1 To 10 ml hybridization buffer add 292.5 mg NaCl (final concentration of 0.5 M).① [1]

2 Add 0.5 g blocking agent and immediately mix to a fine suspension.

3 Continue to mix slowly on a magnetic stirrer for 1 h at RT and then at 42°C for 30–60 min with only occasional stirring.

4 Place Northern/Southern blot into a hybridization vessel (bag/chamber/bottle). If the blot is less than 20 cm^2 then add 0.25 ml/cm^2 of the

Continued overleaf

Notes

① A concentration of 0.5 M NaCl generally gives acceptable results. Alter to optimize if necessary.

② Add buffer whilst still at 42°C.

③ High background will result from insufficient circulation of buffer.

④ Avoid touching pipette tip on membrane when adding probe.

⑤ Reduce SSC concentration of primary wash solution to 0.5× to obtain greater stringency. Avoid longer wash times as sensitivity is lost.

prepared hybridization buffer. If the blot is more than 20 cm^2 then add 0.125 ml/cm^2 of the prepared hybridization buffer.②

5 Seal the hybridization vessel ensuring that the membrane is immersed or free of air bubbles (bag system). Allow the membrane to prehybridize for 1 h at 42°C in an orbital incubator at ≥100 r.p.m.③

6 Add the labeled probe (*Protocol 13a*) to the hybridization fluid and seal the hybridization vessel. Hybridize overnight with gentle agitation.④

7 Preheat the primary wash buffer of choice to 42°C (with urea) or 55°C (without urea). Place membranes in 2–5 ml/cm^2 of primary wash solution and wash at 42°C (urea wash) or 55°C with gentle agitation for 20 min. Repeat the wash.⑤

8 Place the membranes in 20 ml secondary wash solution (20× SSC) and gently agitate for 5 min at RT and repeat.

Pause point

1 Can be kept for up to 3 months at 2–8°C.

Reagents

Reagent supplied with ECL kit (Amersham International):
Detection reagents 1 and 2 (mix a 1:1 solution)

Equipment

Container trays
Saranwrap
X-ray film (e.g. hyperfilm-ECL)

Technique

1 Mix 0.065 ml/cm^2 (membrane) of detection reagents 1 and 2.[1]

2 Drain the excess secondary wash buffer (*Protocol 13b*) from the blot and place in a fresh tray and add the detection reagent mixture to the face bearing the fixed DNA/RNA. Incubate for 1 min at RT.(1)

3 Drain off the excess detection reagents and place the membrane between two sheets of Saranwrap.

4 Switch off the lights in the dark room, remove a film sheet from the stock container and cut off a corner for orientation. Place the sheet over the membrane orientating the cut corner to the top right and hold the film in place for 10 sec.(2)

5 Immediately develop the film and assess exposure time necessary to avoid over- and under-exposure. Repeat the exposure until the record of signal intensity is satisfactory.(3)

Notes

(1) Excess secondary wash buffer can be removed by laying membranes on Whatman™ 3MM (DNA face up) for a few seconds.

(2) Avoid any contact of film and moisture.

(3) Low target applications should be exposed for 30 min in an autoradiographic cassette.

Pause point

[1] This solution may be stored for up to 30 min on ice.

V   GEL ELECTROPHORESIS OF RNA

In the late 1960s Loening and his co-workers developed high resolution gel electrophoretic methods for separating RNA which provide much higher resolution than can ever be achieved by centrifugation. Hence it is not surprising that gel electrophoresis has become the method of choice for determining the size of RNA. For this reason a wide range of analytical and micro-techniques have been devised around this method which allow extensive analysis of small quantities of RNA [1].

Factors influencing molecular migration of RNA

The basis of the separation of RNA by gel electrophoresis has much in common with centrifugal separations, in that molecules of RNA are separated on the basis of their size and conformation. Molecular size is the main determinant of migration speed, where smaller molecules move through the matrix faster than larger molecules. Compact molecules also migrate faster than extended molecules of the same size. However, the same problems of RNA aggregation as found in centrifugal preparations are also encountered. It is therefore necessary to use denaturing gels to determine the real size of RNA in the absence of conformational factors, aggregation and nicks in the RNA. The choice of denaturing conditions depends on the nature of the gel and the electrophoresis conditions; some of the denaturants used are given in *Table 4*. A fact worthy of consideration during experimental design is that nonlinear RNA migrates more slowly than would be expected from its size as it cannot pass through the pores as easily as linear RNA can.

Gel matrices

Polyacrylamide gels provide good separations and are convenient to use for almost all except the largest RNAs [2,3]. Gel recipes suitable for separating RNA are given in *Table 5*. The actual choice of gel concentration depends on the size of the RNA. For the largest types of RNA the acrylamide concentration required is so low that the gels become fragile and difficult to handle; in this case the gels can be strengthened by the addition of 0.5% (w/v) of agarose [2]; these gels are commonly referred to as composite gels. However, agarose gels are far simpler to prepare and handle and are therefore favored for routine analyses. Tris-borate (TBE) is the standard electrophoresis buffer used in many molecular biology laboratories.

Table 4. Types of denaturing conditions used for gel electrophoresis

Denaturant	Concentration	Type of gel	Comments
Dimethylsulfoxide (DMSO)	50–90%	Agarose	Toxic
Glyoxal	10–30%	Agarose	Usually used in combination with DMSO in denaturing agarose gel electrophoresis
Formaldehyde	3% (w/v)	Agarose and polyacrylamide	Toxic. Usually used in denaturing agarose gel electrophoresis.
Formamide	50–98%	Agarose and polyacrylamide	Toxic. Usually used in loading buffers
Heat	60–80°C	Agarose and polyacrylamide	Usually used in concert with other denaturants for initial denaturation
Methyl mercuric hydroxide	3–5 mM	Agarose	Highly toxic. Reacts with imino bonds of uracil and guanine in RNA preventing secondary structure formation. Used in denaturing agarose electrophoresis
Sodium iodoacetate	10 mM	Agarose	Used in the denaturing glyoxal/DMSO system for northern blotting
Urea	6–8 M	Polyacrylamide	Used in denaturing polyacrylamide gel electrophoresis (PAGE) protocols

Agarose

Agarose is also used, not just to strengthen gels but to create a sieving matrix similar to acrylamide matrices. These gels are also used for separating nucleic acid and have the advantage that they are much easier to prepare. Agarose is melted into the buffer of choice and usually poured as a slab gel. The choice of gel matrix and gel concentration again depends on the nucleic acid being separated and the conditions to be used during separation.

Table 5. Acrylamide mixes for preparing polyacrylamide gels

	Final polyacrylamide concentration (%)					
	2.5	3.0	3.5	4.0	5.0	10.0
Acrylamide stock soln (ml)[a]	5.0	6.25	8.0	8.0	10.0	20.0
5× TBE[b] (ml)	6.0	6.0	6.0	6.0	6.0	6.0
Water	19.0	17.75	16.0	14.0	14.0	4.0
Final volume	30.0	30.0	30.0	30.0	30.0	30.0

[a] Acrylamide stock: 30 g acrylamide, 1.5 g bisacrylamide made up to 200 ml with distilled water.
[b] TBE, Tris-borate-EDTA electrophoresis buffer (see *Appendix B*).

Polyacrylamide

Polyacrylamide gel matrices are formed by polymerizing monomers of acrylamide with monomers of a cross-linking agent. The most commonly used cross-linking agent is *N,N′*-methylene-*bis*-acrylamide, referred to as 'Bis' for short. A three-dimensional matrix is formed where the concentration of acrylamide determines the average length of the polymer chains and the Bis concentration defines the extent of cross-linking. The ratio of these monomers defines the eventual pore size, density, elasticity and mechanical strength of the gel. Other cross-linking agents are used for special purposes usually when gel segments are to be solubilized. Agents such as *N,N′-bis*-acrylylcystamine (BAC) or diallyltartardiamide (DADT) allow gel solubilization in 2-mercaptoethanol and 2% periodic acid, respectively [1].

Polymerization is normally initiated by ammonium persulfate (APS) and the reaction is accelerated or catalyzed by *N,N,N′,N′*-tetramethylethylenediamine (TEMED). This is the favored initiation process as it gives repeatable homogeneous pore sizes throughout the gel. Another polymerization method utilizes riboflavin. Riboflavin triggers polymerization when the gel is exposed to ultraviolet illumination. This allows polymerization to be initiated after the gel is poured, avoiding premature polymerization. Polymerization is inhibited by dissolved oxygen in the gel mixture, making it necessary to de-gas the gel mix before polymerization initiation. De-gassing is routinely performed with water-suction pumps but rotary vacuum pumps can also be used.

The concentration of the gel matrix need not be homogeneous, indeed gradient gels with higher concentrations at the base than at the top often afford a better separation. Gradient gels are always acrylamide based and can be made by manually pouring different concentrations into tubes or plates, or by use of a gradient maker (*Figure 1*).

Polyacrylamide separations can be carried out in two dimensions. Two-dimensional electrophoresis is used to separate complex nucleic acid mixtures that cannot be resolved in a one-dimensional gel. A shift in electrophoresis conditions occurs between the first and second dimensions which can be a polyacrylamide matrix concentration change, a shift from non-denaturing to denaturing conditions or a shift in pH combined with denaturation and concentration changes.

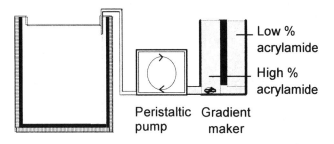

Figure 1. Preparation of a continuous gradient gel using a gradient maker. Acrylamide stock solutions are mixed together to form a continuous gradient.

Gel electrophoresis of RNA

Composite

The addition of 0.5% (w/v) agarose to low-percentage polyacrylamide gels forms a mechanically stable yet very porous gel. Gels composed of 3.0% (w/v) polyacrylamide and 0.5% (w/v) agarose can be handled easily, and gels with even lower polyacrylamide concentrations can be used if necessary. The high resolution provided by these gels permits the separation of large macromolecules which differ, for example, only in conformation or by the presence of a single ribosomal protein. The separation is on the basis of molecular sieving and the extent of separation is determined primarily by the structure (size and shape) of the particle rather than by its charge.

Two-dimensional gel electrophoresis

Two-dimensional electrophoresis is used to separate complex nucleic acid mixtures that cannot be resolved in a one-dimensional gel. Polyacrylamide gels are used in this method, where a shift in electrophoresis conditions occurs between the first and second dimensions. The shift in separating conditions can be achieved by a polyacrylamide matrix concentration change, a shift from nondenaturing to denaturing conditions or a shift in pH combined with denaturation and gel concentration changes.

A shift in polyacrylamide concentration allows RNA molecules with different conformations to be separated. The acrylamide concentration in this method in the second dimension is twice that of the first dimension. A concentration shift of this type often gives well-defined, tight banding and can be used to separate RNAs that vary in size of 80–400 nt [3]. A shift between nondenaturing and denaturing conditions separates molecules that have hidden nicks in their structure. Urea is used as the denaturation agent in these shift experiments at concentrations varying from 4 to 8 M. RNA molecules containing nicks can remain as single units due to the secondary and tertiary interactions. In denaturing conditions these interactions are prevented and separation occurs. A change of pH from acidic pH 3.3 in a denaturing gel to nondenaturing conditions in higher acrylamide concentration gel at pH 8.0–8.3 forms the basis of the third method. Here initial separation in an acidic environment separates the molecules on a basis of base composition. Nucleotide bases (A, C, G, U) contribute

Table 6. Selected two-dimensional systems used for separating RNA molecules

Shift type	First dimension		Second dimension		Type of RNA separated	Reference
	Gel conc. (%)	Conditions	Gel conc. (%)	Conditions		
Polyacrylamide concentration	10	Nondenaturing at neutral pH	20	Nondenaturing at neutral pH	Small RNAs	5
	10.4	Denaturing (4 M urea) at neutral pH	20.8	Denaturing (4 M urea) at neutral pH	tRNAs and precursors	6
Denaturation	6	Nondenaturing at neutral pH	6	Denaturing (5 M urea) at neutral pH	mRNAs	7
	12.5	Nondenaturing at neutral pH	12.5	Denaturing (8 M urea) at neutral pH	5S RNA fragments	8
	15–16	Denaturing (6–7 M urea) at neutral pH	16	Nondenaturing at neutral pH	tRNAs	9
pH denaturation and poly-acrylamide concentration	10.3	pH <4.5 in denaturing conditions (6 M urea)	20.6	Nondenaturing at neutral pH	Viral RNA fragments	10
	10.3	pH <4.5 in 6 M urea	20.6	Nondenaturing at neutral pH	Viral RNA oligonucleotides	11

different charges to the net charge of a RNA molecule at pH 3.3. Therefore molecules of discrete compositions have different mobilities in electric fields under these conditions. This separation factor is rarely sufficient to differentiate between molecules in isolation but can be combined with concentration and denaturation shifts [4]. Some examples of the three main types of shift are detailed in *Table 6*.

Gel concentrations are defined by the sizes of RNA to be separated. For example mRNA varying in size from 800 bases to 5 kb can be separated on a 6% polyacrylamide gel whilst smaller tRNA can be separated on a 15% polyacrylamide gel.

Table 7. Commercially available molecular weight markers

Type of marker	Source
RNA markers	BDH, BHM, GBL, HSI, IBI, PMB, PML, SCC
DNA markers	APP, BHM, BRL, GBL, HSI, IBI, NBL, PMB, PML, SCC, STG
Pulse field markers	BHM, BRL, FIL, PMB, PML, SCC, STG
Oligonucleotide sizing markers	PMB

[a]See Appendix C: Suppliers.

Molecular markers

Calibration of the gels is achieved by running nucleic acid size markers alongside the sample in the same gel. By comparison of the migration of size markers and that of samples, the sizes of the nucleic acid sample components can be determined. Molecular markers are commercially available for the analysis of DNA and RNA. *Table 7* lists selected 'ladders' and *Figure 2* shows some typical separations.

Apparatus used in gel electrophoresis

Polyacrylamide gel electrophoresis (PAGE) may be carried out in tube or slab gels. Tube gels are used to separate single samples per gel but are now rarely used as slab gels are easier to handle. The use of tube gels can be an advantage when cross-contamination of nucleic acid samples is to be avoided as they can be dealt with individually. The internal diameter of the glass tubes ranges from 6 to 9 mm and they are usually 8–12 cm in length.

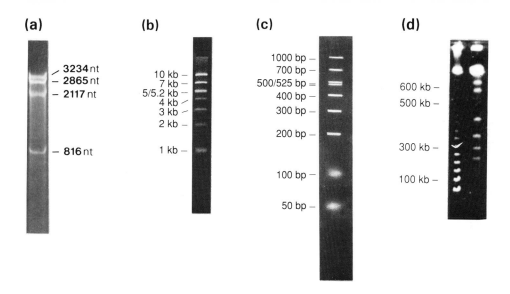

Figure 2. Molecular size markers. (a) Brome Mosaic Virus (BMV) RNA markers (Promega) in a denaturing 1.5% agarose gel (5 V/cm); (b) A 1–10 kb DNA marker in 1% SeaKem® Gold agarose in 1× TAE buffer, 5 V/cm; (c) a 50–1000 bp DNA marker in 4% NuSieve® 3:1 agarose in 1× TBE buffer, 5 V/cm; (d) lambda ladders and *Saccharomyces cerevisiae* chromosomal DNA using a pulsed field CHEF system, 1% SeaKem® LE agarose gel, 6 V/cm, 40 sec, 20 h. Figure reproduced with permission from Flowgen and FMC BioProducts.

Gel electrophoresis of RNA

Polyacrylamide slab gels are frequently more convenient than tube gels for nucleic acid separation. Eight or more samples can be separated in the same gel matrix, being subject to exactly the same running conditions. The use of slab gel allows samples to be aligned and compared with a much greater accuracy than can be achieved by tube gel electrophoresis. The use of slab gels for preparative separations is favored as thicker gels can be used whilst large-bore tube gels often have heat dissipation problems. The electrophoresis tank (*Figure 3*) has upper and lower reservoirs connected only when a gel joins them. This ensures that any electrophoretic flow passes through the gel and does not dissipate into the running buffer. Gels are cast between two glass plates separated by 1–3 mm-thick plastic strips. Plates are cleaned before each use with laboratory detergent, rinsed with distilled water and then with a 1:1 (v/v) ethanol–ether solution and dried with tissue. Plates are sealed with rubber tubing and molten 0.5% (w/v) agarose before gel mixes are poured. Gels usually set in under 30 minutes but it is normal procedure to leave then overnight to ensure complete matrix formation.

Submerged or 'submarine' slab gels in which the gel is completely immersed (*Figure 4*) are popular because they are very easy to prepare, load and analyze. Agarose slab gels allow separation of 10 or more RNA samples on the same gel with only a slightly lower resolution than polyacrylamide slab gels. The advantages of the use of agarose far outweigh this slight loss of resolution. Agarose is a nontoxic compound, less expensive than polyacrylamide and needs no cross-linking agents or initiators.

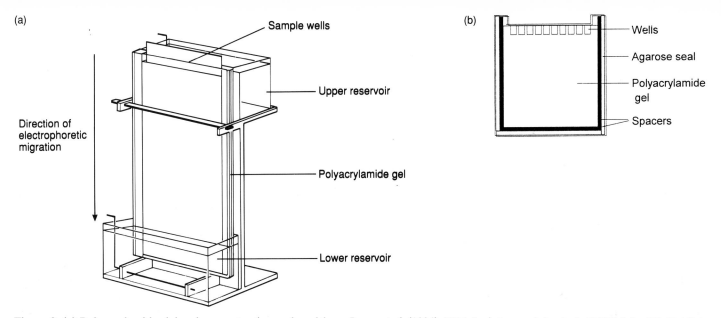

Figure 3. (a) Polyacrylamide slab gel apparatus (reproduced from Jones *et al.* (1994) *RNA Isolation and Analysis*, BIOS Scientific Publishers, Oxford). (b) Glass plates used to form polyacrylamide gel.

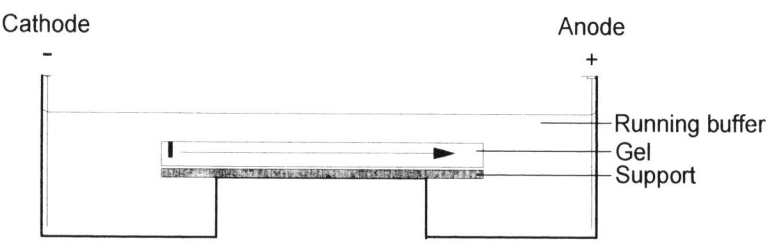

Figure 4. Agarose slab gel apparatus. Used in routine analysis, for example restriction analysis of digests of nucleic acids.

Methods available

Nondenaturing agarose gel electrophoresis (see *Protocol 14*)
Agarose gels are the easiest gels to prepare and are usually run in a horizontal electrophoresis tank. Routine analyses of nucleic acid size during molecular manipulation can be carried out quickly by running DNA/RNA on a gel.
Problems: Ethidium bromide used in the gel matrix to stain the nucleic acid is carcinogenic so gloves must be worn.

Denaturing agarose gel electrophoresis (see *Protocol 15*)
Denaturing agarose gel electrophoresis is the standard method of analysis of specific RNA species where RNA is separated and the gel blotted prior to northern analysis (*Protocol 29*). Two systems are standardly used to separate the RNA, the formaldehyde method or

References

1. Grierson, D. (1990) in *Gel Electrophoresis of Nucleic Acids: a Practical Approach*, 2nd edn. (D. Rickwood and B.D. Hames, eds.) p. 1. IRL Press, Oxford.
2. Peacock, A.C. and Dingman, C.W. (1968) *Biochemistry*, **7**:688.
3. Dahlberg, A.E. and Grabowski, P.J. (1990) in *Gel Electrophoresis of Nucleic Acids: a Practical Approach*, 2nd edn. (D. Rickwood and B.D. Hames, eds.) p. 275. IRL Press, Oxford.
4. De Wachter, R., Maniloff, J. and Fiers, W. (1990) in *Gel Electrophoresis of Nucleic Acids: a Practical Approach*, 2nd edn. (D. Rickwood and B.D. Hames, eds.) p. 151. IRL Press, Oxford.
5. Ikemura, T. and Dahlberg, J.E. (1973) *J. Biol. Chem.* **248**:5024.

the DMSO/glyoxal method. The formaldehyde method is presented in this text and the glyoxal method may be found in ref. 1 of Chapter II. The formaldehyde method is easier to carry out and produces bands of equal definition.

Problems: Formaldehyde is toxic so electrophoresis must be carried out in a fume cabinet.

One-dimensional denaturing slab gel electrophoresis

(see *Protocol 16*)

Polyacrylamide gel electrophoresis is used for high resolution analysis of nucleic acid size. It is harder to carry out than agarose gel electrophoresis but yields a more accurate assessment of nucleic acid size.

Problems: Polyacrylamide is neurotoxic in powder and liquid form so gloves must be worn. Low concentration gels are also fragile and can split easily.

Two-dimensional polyacrylamide gel electrophoresis

(see *Protocol 17*)

Two-dimensional electrophoresis is used to separate complex nucleic acid mixtures that cannot be resolved in a one-dimensional gel. Polyacrylamide gels are used in this method, where a shift in electrophoresis conditions occurs between the first and second dimensions. The shift in separating conditions can be achieved by polyacrylamide matrix concentration change, a shift from nondenaturing to denaturing conditions or a shift in pH combined with denaturation and con-

6. Fradin, A., Gruhl, H. and Feldman, H. (1975) *FEBS Lett.* **50**:185.
7. Burckhardt, J. and Birstiel, M.L. (1978) *J. Mol. Biol.* **188**:61.
8 Vigne, R. and Jordan, B.R. (1971) *Biochemie*, **53**:981.
9 Stein, M. and Varricchio, F. (1974) *Anal. Biochem.* **61**:112.
10. De Wachter, R. and Fiers, W. (1972) *Anal. Biochem.* **49**:184.
11. Billeter, M.A., Parsons, J.T. and Coffin, J.M. (1974) *Proc. Natl Acad. Sci. USA*, **71**:3560.
12. Berk, A.J. and Sharp, P.A. (1977) *Cell*, **12**:721.
13. Favaloro, J., Treisman, R. and Kamen, R. (1980) *Meth. Enzymol.* **56**:718.
14. Humphrey, T., Christofori, G., Lucijanic, V. and Keller, W. (1987) *EMBO J.* **6**:4159.
15. Cotton, R.G.H. (1993) *Mutation Res.* **285**:125.

Protocols provided

centration changes. The first and second dimensions of two-dimensional gel electrophoresis are physically separate. Samples are electrophoresed through a gel that constitutes the first dimension, a gel band is cut from the gel and placed into a gel mould around which the second-dimension gel is poured. Electrophoresis in the second dimension allows fragments separated in the excised band to migrate under different conditions through the second gel. The first-dimension gel segment can also be treated with restriction endonucleases in detailed analyses.

Problems: Patterns obtained from two-dimensional gel electrophoresis may be too complicated to analyze, particularly if a restriction endonuclease step is incorporated into the method. Combining electrophoresis with a method of probing for a specific DNA/RNA fragment can simplify analysis of the two-dimensional separation.

Nuclease S1 analysis (see Protocol 18)

This procedure can be used to map RNA termini [12], to quantitate specific RNA species or to locate splice junctions [13]. The method relies on a hybridization of RNA and DNA which is favored over DNA:DNA hybridization in this procedure. Nuclease S1 then hydrolyses DNA that has not hybridized leaving only protected hybrids. One of the members of the hybrid is usually radiolabeled and is subsequently analyzed using gel electrophoresis to assess its size.

Problems: RNA precipitated with ethanol or dried in a desiccator prior to nuclease S1 analysis can cause problems in the hybridization step by incomplete dissolution. If this occurs, repeat the procedure but concentrate the RNA sample in a rotary evaporator rather than by ethanol precipitation until just dry and redissolve in hybridization solution (*Protocol 18*).

Primer extension analysis (see *Protocol 19*)
Primer extension analysis is the standard method used to determine transcription start sites of mRNA species in relation to the gene. A primer downstream of the transcription start site is used to hybridize to the mRNA allowing the enzyme reverse transcriptase to copy the mRNA in a $5' \rightarrow 3'$ direction (towards the start of the mRNA) and will do so until it has no template to copy. The size of the extended primer is then assessed using gel electrophoresis and the transcription start site determined or the extended primer may be sequenced using the Maxam–Gilbert method.

Problems: The primer site should be within 100 bases of the expected RNA termini, as extension for more than 100 bases leads to the enzyme dropping-off early. Multiple signals can occur if the RNA contains high levels of secondary structure that is not completely unfolded at the start of the procedure.

Gel electrophoresis of RNA

Protocol 14. Nondenaturing agarose gel electrophoresis of RNA (or DNA)

Reagents

5 mg/ml Ethidium bromide (EtBr) solution⚠

Molecular weight markers

0.1 M NaOH⚠

Sample buffer (50% (w/v) glycerol, 1 mM EDTA (pH 8.0), 0.25% (w/v) bromophenol blue, 0.25% (w/v) xylene cyanol FF)

SeaKem LE agarose or equivalent

10× TBE (0.9 M Tris base, 0.9 M boric acid, 25 mM EDTA)

Equipment

Camera

Microwave oven or hotplate

Submarine gel electrophoresis apparatus

UV light box⚠

Technique

1 Prior to electrophoresis soak the electrophoresis tank, gel tray and comb in 0.1 M NaOH for at least 6 h (omit for DNA gels).①

2 Remove the gel tray and comb from the tank and rinse with autoclaved double-distilled water. Dry the edges of the tray with tissue paper and seal with tape.

3 Add 1.5 g research grade agarose (e.g. SeaKem LE agarose) to 100 ml 1× TBE electrophoresis buffer. Melt agarose in a Pyrex vessel by boiling (hotplate or microwave) for 2 min. Ensure completion by gentle swirling of the solution, checking for the presence of undissolved agarose.

Notes

① NaOH should remove any RNase activity present on the equipment.

② The gel is cooled in order to avoid cracking of the gel tray upon pouring.

③ Total amount (μg) of RNA to be loaded is limited by well volume, e.g. a 50 μl well can be loaded with 50 μg. Large amounts of RNA should be electrophoresed at 30 mA for 2–3 h to prevent streaking.

4 Cool the gel solution by running cold water over the vessel containing the gel until it is at approximately 50°C (just possible to be held comfortably in the hand). Add 10 µl 5 mg/ml EtBr to the gel mixture and mix by gentle swirling.②①

5 Pour the gel solution into the gel tray until the solution touches the top of the teeth of the comb. Leave the gel to set for 30 min at RT.

6 Rinse the electrophoresis tank with autoclaved double-distilled water and fill tank with 1× TBE to a level that will cover the gel.

7 Remove the comb from the gel and place the gel into the tank.

8 Total RNA of between 0.1–1.0 µg/µl can be loaded on to the gel with 1/10 vol. of the sample buffer. Electrophorese the RNA samples together with any required molecular weight markers at 20–25°C for 1 h at 80 V.③

9 Visualize the RNA (/DNA) bands by placing the gel on a UV light box and record the image either using an image capture device (e.g. Mitsubishi video copy processor) or by Polaroid photography (typical settings: f4.5, 1/30–1/4 sec using Polaroid Polablue (135 mm) film).

Pause point

[1] Gel can be left in the tank, submerged in buffer, for up to 24 h prior to use.

Protocol 14. Nondenaturing agarose gel electrophoresis of RNA (or DNA)

Protocol 15. Denaturing agarose gel electrophoresis of RNA

Reagents

Denaturation solution (0.7× MOPS buffer, 9.2% formaldehyde)⚠
5 mg/ml Ethidium bromide (EtBr) solution⚠
38% (w/v) Formaldehyde⚠
Molecular weight markers
10× MOPS (0.2 M MOPS (3-(N-morphino)propanesulfonic acid), 50 mM sodium acetate, 5 mM ethylenediaminetetra-acetic acid (EDTA))
0.1 M NaOH⚠
1× Running buffer (1× MOPS buffer, 3% (w/v) formaldehyde)⚠

Sample buffer (50% (w/v) glycerol, 1 mM EDTA (pH 8.0), 0.25% (w/v) bromophenol blue, 0.25% (w/v) xylene cyanol FF)
SeaKem LE agarose or equivalent

Equipment

Camera
Fume cabinet
Microwave oven or hotplate
Submarine gel electrophoresis apparatus
UV light box⚠
Water baths at 60°C and 65°C

Technique

1 Prior to electrophoresis soak the electrophoresis tank, gel tray and comb in 0.1 M NaOH for at least 6 h. Remove the gel tray and comb from the tank and rinse with autoclaved double-distilled water. Dry the edges of the tray with tissue paper and seal with tape.①

2 Add 1.2 g research-grade agarose (e.g. SeaKem LE agarose) to 72 ml sterile double-distilled water. Melt the agarose by boiling in a Pyrex vessel (using hotplate or microwave) for 2 min. Ensure completion by gentle swirling of the solution, checking for the presence of undissolved agarose.

Notes

① NaOH should remove any RNase activity present on the equipment.
② The gel is cooled in order to avoid cracking of the gel tray upon pouring.

3 Cool the gel solution by placing the bottle containing the gel in a 60°C water bath for 10 min, swirling each minute. Add 10 μl of 5 mg/ml EtBr to the gel mixture and mix by gentle swirling. In a fume cabinet, add 10 ml 10× MOPS and 18 ml 38% formaldehyde solution, mixing by gentle swirling.②

4 Pour the gel solution into the gel tray until the solution touches the top of the teeth of the comb. Leave the gel to set for 30 min at RT in the fume cabinet.

5 Rinse the electrophoresis tank with autoclaved double-distilled water. Fill the tank with 1× running buffer to a level that will cover the gel. Remove the comb from the gel and place the gel into the tank.

6 Total RNA (5–25 μg) is mixed with an equal volume of denaturation solution and placed at 65°C for 2 min. (Treat molecular markers as samples.) Add 1/10 vol. of sample buffer and load the samples immediately. Electrophorese the RNA samples together with any required molecular weight markers at 20–25°C for 1 h at 80 V.

7 Visualize the RNA bands by placing the gel on a UV light box. Record the image either using an image capture device (e.g. Mitsubishi video copy processor) or by Polaroid photography (typical settings: f4.5, 1/30–1/4 sec using Polaroid Polablue (135 mm) film).

Protocol 16. One-dimensional denaturing slab gel electrophoresis

Reagents

40% Acrylamide stock solution (38% acrylamide, 2% methylene bisacrylamide in water), store at 4°C△

Agarose

10% Ammonium persulfate (APS)△

Ethanol:ether (1:1)△

Ethidium bromide (EtBr) solution (5 mg/ml)△

Loading buffer (50% glycerol, 1 mM ethylenediaminetetra-acetic acid (EDTA) pH 8.0), 0.25% bromophenol blue, 0.25% xylene cyanol FF in deionized formamide)

10× TBE (0.9 M Tris base, 0.9 M boric acid, 20 mM EDTA)

N,N,N',N'-tetramethylethylenediamine (TEMED)△

Urea

Equipment

Electrophoresis tank

Gel plates

Gel spacers (1–3 mm thick)

Metal bulldog clips

Microwave

Nalgene filter (0.2 µm pore size)

Power pack

Sealable gel container

Syringe

UV transilluminator△

Technique

1 Clean the glass gel plates with detergent, rinse with distilled water and rub over with ethanol:ether (1:1 v/v). Place the plastic spacers between the plates on the left, right and bottom sides leaving 5 mm space from the edge. Clamp the plates together with two metal bulldog clips on the left, right and bottom sides.

Notes

① This is a 4% denaturing polyacrylamide gel mix. On hot days chill gel mix on ice to avoid premature polymerization.

② Persistent bubbles can be removed during pouring by asking a colleague to tap at the gel border with a heavy pen.

2 Melt 0.1 g agarose in 10 ml water in a microwave oven and apply around the spacers with a Pasteur pipette, creating an agarose seal.

3 In a glass beaker mix 5 ml 40% acrylamide stock solution, 21 g urea, 5 ml 10× TBE and 10 ml sterile, double-distilled water. Dissolve urea on a magnetic stirrer and make final volume to 50 ml with sterile, double-distilled water. Filter solution through a 0.2-μm-pore Nalgene filter.①

4 Add 30 μl TEMED and 300 μl of 10% APS, mix and quickly pour between the gel plates, avoiding bubble formation.②

5 After the gel has polymerized, remove the clips and the bottom spacer and place into the gel tank. Fill the upper and lower reservoirs with 1× TBE to a height of 2 cm above the wells and base of gel. Squirt 1× TBE into the wells to remove any urea that has diffused from the gel.

6 To 0.5–30 μg RNA add 1/10 vol. loading buffer and boil the RNA for 2 min. Collect by brief centrifugation. Load the samples into wells and electrophorese at 5 V/cm until dark dye line (bromophenol blue) is 3/4 the length of the gel from the wells.③

7 Remove the gel plates from the tank and place on a bench surface. Place a spatula between the plates and twist gently to part the plates, leaving the gel on one plate. The gel may then be stained, electroblotted (*Protocol 27*) or certain RNA species recovered by elution (*Protocol 33*).

③ Load samples in to the base of the well slowly for maximum volume loadings.

Continued overleaf

Protocol 16. One-dimensional denaturing slab gel electrophoresis

8 Staining: place the gel plate bearing the gel at a slant into a container of electrophoresis buffer containing 5 µg/ml EtBr (1 µl EtBr/ml buffer). Fill a syringe with electrophoresis buffer and squirt between the gel and the plate to release the gel from the plate into a container. Leave the gel immersed at RT for 30 min.

9 Destain by carefully decanting EtBr solution off and replace with electrophoresis buffer for 5 min. Visualize the RNA bands by placing the gel on a UV transilluminator.

10 Record the image either using an image capture device (e.g. Mitsubishi video copy processor) or by Polaroid photography (typical settings: f4.5, 1/30–1/4 sec using Polaroid Polablue (135 mm) film).

Protocol 17. **Two-dimensional polyacrylamide gel electrophoresis**

Reagents

40% Acrylamide stock solution△ (38% acrylamide, 2%
  methylene bisacrylamide in water), store at 4°C
Agarose
10% Ammonium persulfate (APS)△
Ethanol:ether (1:1)△
Loading buffer (50% glycerol, 1 mM ethylenediaminetetra-acetic
  acid (EDTA) pH 8.0, 0.25% bromophenol blue, 0.25% xylene
  cyanol FF in deionized formamide)
10× TBE (0.9 M Tris base, 0.9 M boric acid, 20 mM EDTA)

N,N,N',N'-tetramethylethylenediamine (TEMED)△
Urea

Equipment

Bulldog clips
Electrophoresis tank
Gel plates
Gel spacers (1–3 mm thick)
Microwave
Nalgene filter (0.2 μm pore size)
Power pack

Technique (ref. 14)

1 Carry out the first dimension *native* separation in a slab gel according to
  Protocol 14.

2 Clean the glass gel plates with detergent, rinse with distilled water and
  rub over with ethanol:ether (1:1 v/v). Place plastic spacers between the
  plates on the left, right and bottom sides leaving 5 mm space from the
  edge.

Continued overleaf

Notes

① The second dimension gel is a 12% denaturing gel.

3 Clamp the plates together with two bulldog clips on the left, right and bottom sides. Melt 0.1 g agarose in 10 ml water in a microwave and apply around the spacers with a Pasteur pipette, creating an agarose seal.

4 In a glass beaker mix 15.63 ml 40% acrylamide stock solution, 21 g urea, 5 ml of 10× TBE and 10 ml sterile double-distilled water. Dissolve urea on a magnetic stirrer and make final volume to 50 ml with sterile double-distilled water. Filter the solution through a 0.2-μm-pore Nalgene filter. Chill on ice.①

5 Add 30 μl TEMED and 300 μl 10% APS, mix and quickly pour between the gel plates, avoiding bubble formation. Save 2 ml of the solution on ice.

6 Push the relevant lane of slab gel into the top of the second dimension gel, ensuring no bubbles are formed. Pour the remaining 2 ml over the first dimension gel and allow polymerization.

7 After the gel has polymerized, remove the clips and bottom spacer and place into the gel tank. Fill upper and lower reservoir with 1× TBE to a height of 2 cm above the wells and the base of gel. Squirt 1× TBE into the wells to remove any urea that has diffused from the gel.

8 Electrophorese the gel at 5 V/cm for 4–8 h or until the bromophenol blue is 3/4 the length of the gel from the wells.

9 Remove the gel plates from the tank and place them on a bench surface. Place a spatula between the plates and twist gently to part the plates, leaving the gel on one plate. The gel may then be stained (*Protocol 16*), electroblotted (*Protocol 27*), autoradiographed (*Protocol 31*) or RNA species recovered by elution (*Protocol 33*).

Protocol 18. Nuclease S1 analysis

Reagents

Ethanol⚠

5′-End labeled DNA probe

Hybridization buffer (80% (v/v) deionized formamide, 40 mM piperazine-1,4-*bis*(2-ethanesulfonic acid) (PIPES)-KOH pH 6.4⚠, 1 mM ethylenediaminetetra-acetic acid (EDTA), 0.4 M NaCl)

Loading solution (95% formamide, 20% sucrose, 0.25% bromophenol blue, 0.13% xylene cyanol, 0.1× TBE, 0.1% SDS)⚠

Nuclease S1 mapping buffer (0.28 M NaCl, 0.05 M sodium acetate, 4.5 mM $ZnSO_4$ and 100–1000 units/ml nuclease S1)

Phenol:chloroform (1:1)⚠

Sample RNA (0.5–250 μg)

3 M Sodium acetate, pH 5.2

Stop mixture (4 M ammonium acetate, 50 mM EDTA, 50 μg/ml yeast tRNA)

Equipment

Microcentrifuge

Water bath at hybridization temperature

Water bath at 85°C

Technique

1 5′-End label single stranded DNA probes according to *Protocol 7*.

2 Mix the DNA probe and RNA in a ratio suitable to the RNA abundance① and probe size (specific activity)② (e.g.0.5–250 μg RNA and 0.1–1.0 μg DNA probe).

3 Precipitate DNA and RNA by adding 0.1 vol. 3 M sodium acetate and 2.5 vols ethanol. Place at –20°C for 20 min and collect the precipitate by

Notes

① If RNA is low abundance then use approx. 250 μg of sample RNA. When RNA target is abundant then use ≥ 0.5 μg of sample RNA.

② Large DNA probes (≥ 5 kb): use 1 μg of probe. When using small probes use 0.1 μg of DNA.

centrifugation at 15000 g for 15 min at 4°C. Gently wash the pellet with ice-cold 70% ethanol. Dry the pellet at RT.

4 Dissolve the pellet in 30 µl hybridization buffer and heat denature at 85°C for 10 min. Quickly collect the condensation by pulsing in a microcentrifuge and transfer to a water bath at the hybridization temperature③ for 12–16 h.

5 Add 300 µl ice cold nuclease S1 mapping buffer and place at 37°C for 30 min. Terminate the reaction by adding 80 µl stop mixture.

6 Add 330 µl phenol:chloroform, mix and separate the phases by centrifugation at 15000 g. Recover the upper aqueous layer and precipitate the hybrid by addition of 660 µl ice-cold ethanol.

7 Collect the precipitate by centrifugation at 15000 g and 4°C for 15 min. Redissolve the pellet in 10 µl of loading solution and analyze on a denaturing polyacrylamide gel according to *Protocol 16* (acrylamide concentration depends on the size of probe).

③ Hybridization temperature is defined by the G:C content of the hybrid:

%G:C	Temperature
40	48
45	52
50	54
55	59
60	61

Protocol 18. Nuclease S1 analysis

Reagents

10% Acetic acid, 15% methanol⚠

Chloroform⚠

0.1 M Stocks of dATP, dCTP, dGTP, dTTP

1 M Dithiothreitol (DTT)

Ethanol⚠

10× Extension buffer (0.1 M Tris-HCl, pH 8.1, 0.6 M NaCl, 80 mM MgCl$_2$)

HHS buffer (20 mM Hepes-KOH, pH 7.5, 0.25 M sodium acetate, 0.1 mM EDTA)

10× Hybridization buffer (0.1 M Tris-HCl, pH 7.5, 3 M NaCl, 10 mM ethylenediaminetetra-acetic acid (EDTA))

Phenol:chloroform (1:1)⚠

^{32}P-labeled primer (*Protocol 5*)⚠

Reverse transcriptase (20 units/μl)

Equipment

Microcentrifuge

Water bath at hybridization temperature

Water baths at 42°C and 95°C

X-ray film and cassette

Technique (ref. 15)

1 Prepare the reaction mixture: 33 μl 10× extension buffer, 3.3 μl each dNTP stock, 3.3 μl 1 M DTT, 116.5 μl reverse transcriptase (final conc. 5 units/μl), 300 μl water. Prewarm to 42°C.①

2 For each primer analysis reaction, set up the hybridization in a microcentrifuge tube: 1 μl 10× hybridization buffer, 8 μl (1–10 μg) sample RNA and 1 μl (0.5 ng (10^4–10^5 c.p.m.)) ^{32}P-labeled primer. Hybridize at 50–70°C for 10 min.②

3 Quickly transfer the hybridization reactions to 42°C and add 23.3 μl

Notes

① This mix is enough for 10 reactions.

② Hybridization temperature is defined by the G:C content of the hybrid:

%G:C	Temperature
40	48
45	52
50	54
55	59
60	61

pre-warmed extension mixture. Mix by slowly pipetting up and down in a pipette tip. Do not vortex ^{32}P. Leave at 42°C for 20 min.

4 Place the reactions on ice and add 367 µl HHS buffer (final vol. 400 µl).

5 Phenol extract with 400 µl phenol:chloroform, chloroform extract with 400 µl chloroform and precipitate by the addition of 800 µl ice-cold ethanol. Place at −20°C for 30 min and collect by centrifugation at 15 000 g for 10 min.

6 Heat denature samples at 95°C for 5 min prior to loading on to a 10% denaturing polyacrylamide sequencing gel (*Protocol 16*). Run sequencing ladders alongside the samples to assess the size of the primer extension products.

7 Fix the gel by immersion in 10% acetic acid, 15% methanol for 20 min to remove the urea that is present in the gel. Dry the gel on a heated gel drier for 30 min prior to autoradiography.

8 Autoradiography is carried out using Kodak XAR-5 film, although better contrast at the expense of a longer exposure time can be achieved using Kodak Ektascan film.

Protocol 19. Primer extension analysis

V ELECTROPHORESIS OF DNA

DNA protocols

Routine electrophoresis during day-to-day techniques usually involves tasks such as separation of restriction fragments which can quickly be assessed by running on a native gel (please refer to *Protocol 14*: native agarose electrophoresis protocol for RNA). Many analyses involve blotting of these agarose gels which is described in *Protocol 26* and will not be repeated in this chapter.

Electrophoresis of large DNA molecules using the methods detailed in Chapter IV is not always possible. The methods and apparatus have been adapted to allow the separations of megabase DNA molecules. The main technique for carrying this out is pulsed field gel electrophoresis (PFGE) which uses alterations in the electrical field applied to the DNA to separate similar large DNA fragments. Variations on the PFGE apparatus have been derived for specialized experiments (*Figure 5*).

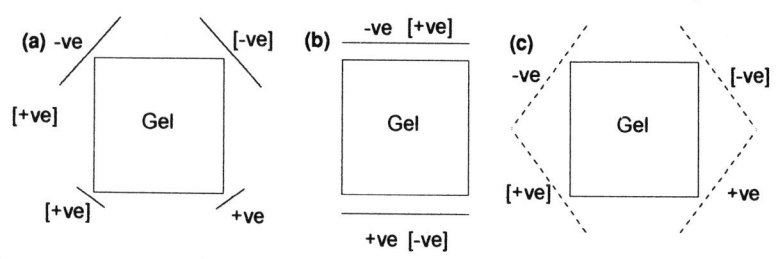

Figure 5. Electrode configurations used in different types of PFGE apparatus. Changes in polarities after the duration of a pulse are shown in brackets. (a) Double inhomogeneous field using long cathodes and short anodes; (b) field inversion gel electrophoresis using a homogeneous electrical field; (c) contour-clamped homogeneous field using a hexagonal electrophoresis chamber with electrodes on four sides of the hexagon. Reproduced from ref. 10 by permission of Oxford University Press.

Methods available

Dideoxy sequencing (see *Protocol 20*)

DNA sequencing is usually approached in one of two ways, either in a large sequencing project or when sequencing is a checking procedure of previous molecular manipulations, for example cloning/mutagenesis. Large sequencing projects now tend to employ a lot of automation of the process but this is not necessary when a clone is to be checked. Automated and semi-automated sequencing procedures will be detailed in a forthcoming book in this series [1]. General laboratory sequencing used to check double-stranded templates (e.g. plasmids) is detailed in *Protocol 20*. PCR sequencing is also now being used to quickly check DNA manipulations but is not as well used as this procedure.

Problems: *During gel pouring* – bubbles often occur when a large sequencing gel is poured particularly with plates that have not been cleaned in the 30 minutes prior to pouring. Bubbles can be avoided by asking a colleague to tap at the gel boundary as you pour the gel to prevent bubble formation. *During electrophoresis* – gel dyes may start to run as a 'smile' which can make the sequence impossible to read even manually. This is usually due to uneven heating during electrophoresis. *After electrophoresis* – sequence located only at the base of the gel can occur for numerous reasons: DNA template concentration, primer concentration, dNTP/ddNTP concentration or the

References

1. Roe, B. (1995) DNA *Isolation and Sequencing*. John Wiley & Sons, Chichester, in press.
2. Brewer, B.J. and Fangman, W.L. (1987) *Cell*, **51**:463.
3. Brewer, B.J. and Fangman, W.L. (1987) *Proc. Natl Acad. Sci. USA*, **91**:3418.
4. Oppenheim, A. (1981) *Nucleic Acids Res.* **9**:6805.
5. Bell, L. and Byers, B. (1983) *Anal. Biochem.* **130**:527.
6. Davidson, F., Simmonds, P., Ferguson, J.C. *et al.* (1995) *J. Gen. Virol.* **75**:1197.
7. Chen, C., Sleper, D.A. and West, C.P. (1995) *Crop Sci.* **35**:720.
8. Jack, P.L., Dimitrijevic, T.A.F. and Mayes, S. (1995) *Theor. Appl. Genet.* **90**:643.
9. McPherson, M.J., Quirke, P. and Taylor, G.R. (eds.) (1991) *PCR: a Practical Approach*. IRL Press, Oxford.
10. Anand, R. and Southern, E.M. (1990) *Gel Electrophoresis of Nucleic Acids: a Practical Approach*, 2nd edn. IRL Press, Oxford.
11. Murphy, G. and Kavanagh, T.A. (1988) *Nucleic Acids Res.* **16**:5198.
12. Brewer, B.J. and Fangman, W.L. (1994) *Proc. Roy. Acad. Sci. USA*, **91**: 3418–3422.
13. Song, K. and Osborn, T.C. (1994) *Plant Mol. Biol.* **26**:1065.
14. Dolzani, L., Tonin, E., Lagatolla, C. and MontiBragadin, C. (1994) *FEMS Microbiol. Lett.* **119**: 167–174.

pH of one or more of the solutions is wrong. The cause is often that a new solution has been used. Isolate the problem empirically by using solutions known to work in the laboratory.

Two-dimensional agarose gel electrophoresis (see *Protocol 21*)
Two-dimensional agarose gel electrophoresis has been used to analyze replication origins [2,3] using the rationale that nonlinear duplex DNA migrates anomalously in agarose gels as compared with a linear molecule of the same size [4,5]. This anomalous behavior is enhanced by increasing either the agarose concentration or the voltage. A combination of a normal separation of DNA in a one-dimensional agarose gel prior to shifting to separation in a higher agarose concentration can therefore be used to investigate the three-dimensional shape of the molecule. *Figure 6* (p. 79) shows some simple cartoons of expected separations of this sort. The procedure has been extended to include a restriction enzyme digestion between the first and second dimensions for further analysis [3] and to map the origin of a replication bubble more exactly.
Problems: The system needs to be optimized for the nucleic acid under analysis altering gel concentration and the voltage during electrophoresis.

Restriction fragment length polymorphism analysis of RNA transcripts using PCR (PCR–RFLP) (see *Protocol 22*)
Restriction fragment length polymorphism (RFLP) analysis is now

15. Ke, S.-H. and Wartell, R.M. (1993) *Nucleic Acids Res.* **21**: 5137–5143.
16. Johnson, P.G. and Beerman, T.A. (1994) *Anal. Biochem.* **220**:103–114.

Protocols provided

the basis of many phylogenetic classification experiments [6–8]. PCR has been adapted to carry out this type of analysis to speed the process up, making diagnostic RFLP analysis a possibility for many laboratories.

Analysis of PCR products: PCR is based upon the copying of the two strands of a DNA template by a DNA polymerase, separating the strands produced and repeating the copying procedure with the larger number of templates available. This process involves the repetitive denaturation by heat which standard DNA polymerases used in other molecular biology procedures cannot withstand. A thermostable DNA polymerase (Taq polymerase), originally isolated from the thermophilic bacteria *Thermus aquaticus*, can survive temperatures which are necessary to denature DNA templates and it is this and similar enzymes that are the cornerstones of PCR. PCR procedures are now being applied to virtually every molecular biology procedure and many of these procedures are described in ref. 9.

Problems: DNA samples must be isolated from only one species as contaminating DNA will lead to confusing patterns after analysis.

PCR ribotyping of bacterial rRNA genes – species/type identification (see *Protocol 23*)

Ribotyping is an analysis usually applied to the number of repeats or intergenic spacing seen in rRNA genes, both of which are highly variable. This procedure, like RFPL analysis, is applied to species and variety/type identification. *Protocol 23* details a procedure where ribotyping has been speeded up by integrating the procedure with PCR.

Problems: Samples must be isolated from only one species as contaminating nucleic acid will lead to confusing patterns after analysis.

Assessment of PCR fidelity using temperature gradient gel electrophoresis (see *Protocol 24*)

The copying process by some Taq polymerases is error prone which leads to slightly different physical properties in the products. Temperature gradient gel electrophoresis (TGGE) can be used to detect small changes (e.g. 1 base) in PCR products prior to cloning and analysis (*Protocol 24*).

Problems: 'Homemade' gel tanks can produce temperature gradients that are not homogeneous leading to strange electrophoretic patterns. Check the temperature gradient with thermocouples periodically.

Pulsed field gel electrophoresis analysis of high molecular weight DNA (see *Protocol 25*)

Pulsed field gel electrophoresis (PFGE) is used to separate very large DNA fragments. This is accomplished by varying the electrode configuration in the apparatus.

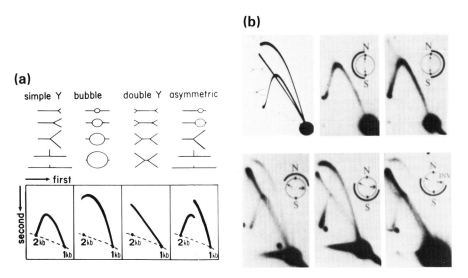

Figure 6. Two-dimensional agarose electrophoresis in replication bubble analysis. (a) Patterns generated by different DNA structures (reproduced from ref. 2 with permission from Cell Press). (b) Patterns generated by nuclease digested DNA, it depicts an expected two-dimensional gel pattern and the two-dimensional separations achieved after electrophoresis. *Protocol 21* describes the basic procedure for analysis of three-dimensional shapes using two-dimensional agarose gel electrophoresis (reproduced from ref. 3 with permission from the National Academy of Sciences of USA).

Electrophoresis of DNA

Protocol 20. Dideoxy sequencing of double-stranded DNA

Reagents

10% Acetic acid, 15% methanol△

[α-^{35}S]dATP (18.5 TBq (500 Ci)/mmol at 370 MBq (10 mCi)/ml)△

Enzyme mix (66 μl 10 mM Tris/0.1 mM EDTA ($T_{10}E_{0.1}$) pH 7.5, 11 μl 0.1 M dithiothreitol (DTT), 4.5 μl label mix (7.5 μM Deaza-GTP, 7.5 μM dCTP, 7.5 μM dCTP in $T_{10}E_{0.1}$))

Fresh 1 M NaOH, 1 mM ethylenediaminetetra-acetic acid (EDTA)△

Reaction buffer (0.2 M Tris-HCl (pH 7.5), 0.1 M MgCl$_2$, 0.25 M NaCl)

Sequenase v2.0 (USB)

Sequencing gel loading buffer (98% formamide, 1 mg/ml xylene cyanol, 1 mg/ml bromophenol blue, 0.01 M EDTA)△

Termination mixes (0.1 mM dATP, dCTP, dGTP, dTTP, 0.01 mM ddNTP in $T_{10}E_{0.1}$)

Equipment

Gel drier
Sepharose CL6B spin columns (*Protocol 3*)
Sequencing gel apparatus
Water bath at 37°C

Technique (ref. 11)

1 Template denaturation: mix 20 μl (2–10 μg) plasmid template with 5 μl fresh 1 M NaOH, 1 mM EDTA and leave at RT for 5 min.

2 Purify the denatured single-stranded DNA by passage through Sepharose CL6B spin columns, prior to primer annealing (see *Protocol 3*).

3 Primer annealing: mix 8 μl template with 1 μl primer (0.3 pmol/μl) and 2 μl reaction buffer and stand for 15 min at 37°C.

Notes

① Siliconized pipette tips should be used as the enzyme adheres to nonsiliconized tips.

② Gels can be loaded quickly by use of one tip for all samples. Use a duck-billed tip, cut off to fit a P10 Gilson, load a sample, wash out tip in lower buffer, remove tip from buffer keeping Gilson plunger down and touch tip to clean tissue paper to remove buffer from tip. Using this technique the tip should not block.

4 Labeling reaction: set up four labeling reactions by aliquoting 2.5 µl of the annealed template solution into lidless microcentrifuge tubes or microtiter plate wells.

5 Add 2 µl enzyme mix, 3 µl (13 units/µl) Sequenase v2.0 (USB) and 5 µl [α-^{35}S]dATP to each tube.①

6 Chain termination: immediately add 2 µl of the termination mixes (A, G, C, T) to the appropriate tubes and stand for 5 min at 37°C.

7 Reaction termination: terminate the reaction with 2 µl sequencing gel loading buffer.

8 Gel electrophoresis: heat denature the samples prior to loading on to the gel by boiling for 2 min and snap cooling to prevent re-annealing. Electrophorese on a 6% acrylamide gel (in 0.5× TBE) at 1600 V, 40 W for 3 h to produce a good separation of the radiolabeled fragments.②

9 Fixation: immerse the gel in 10% acetic acid, 15% methanol for 20 min to remove the urea that is present in the gel.

10 Dry the gel on a heated gel drier for 30 min prior to autoradiography.

11 Autoradiography is carried out overnight using Kodak XAR-5 film. ③

③ Better contrast at the expense of a longer exposure time can be achieved using Kodak Ektascan film.

Protocol 21. Two-dimensional agarose gel electrophoresis

Reagents

Agarose
DNase I (10 units/µl)
DNA to be analyzed at 100 µg/ml
Ethidium bromide (EtBr; 10 mg/ml)⚠
Loading solution (20% (w/v) glycerol, 1 mM EDTA (pH 8.0), 0.25% (w/v) bromophenol blue, 0.25% (w/v) xylene cyanol FF)
Molecular weight markers

1× TBE electrophoresis buffer (0.09 M Tris base, 0.09 M boric acid, 1 mM ethylenediaminetetra-acetic acid (EDTA))

Equipment

Agarose gel electrophoresis equipment (see *Protocol 14*)
UV light box⚠
Water baths at 37°C and 50°C

Technique (ref. 2)

1 Prepare a 0.4% agarose gel solution by mixing 0.4 g agarose in 100 ml 1× TBE and melting on a hotplate or in a microwave. Set up a gel tray and cool the gel to 50°C. Add 5 µl EtBr to the gel solution and pour into the prepared tray (see *Protocol 14*).

2 Randomly nick the plasmid DNA to be analyzed by mixing 10 µl plasmid DNA (100 µg/ml), 1 µl DNase I and 1 µl EtBr. Incubate at 37°C for 10 min.

3 Load the DNA samples with 1/10 vol. loading solution and electrophorese for 15 h at 1 V/cm in 1× TBE. Run size markers alongside (e.g. Lambda digested with *Hind*III and φx174 digested with *Hae*III).

4 Prepare a 1% agarose gel solution by mixing 1 g agarose in 100 ml 1×
TBE and by melting on a hotplate or in a microwave. Set up a gel tray
and cool the gel to 50°C. Add 3 μl EtBr to the gel solution and pour into
the prepared tray (see *Protocol 14*) without inserting the gel comb. Keep
2 ml 1% agarose in a water bath at 50°C.

5 Cut a 1-cm strip from the second dimension gel once it has set ready for
the first dimension gel. Cut the sample lane from the first dimension
agarose gel and lay it along the top of the second dimension gel so that
DNA will pass into the second gel from the gel strip during electrophoresis. Seal the strip in place with the molten agarose kept aside and
allow to set.

6 Electrophorese into the second dimension for 4–8 h at 5 V/cm at 5°C in
1× TBE.

7 Visualize the DNA bands by placing the gel on a UV light box and record
the image either using an image capture device (e.g. Mitsubishi video
copy processor) or by Polaroid photography (typical settings: f4.5,
1/30–1/4 sec using Polaroid Polablue (135 mm) film). Capillary blot the
gel (*Protocol 26*) and analyze by Southern hybridization (*Protocol 28*) if
required.

Protocol 21. Two-dimensional agarose gel electrophoresis

Protocol 22. Restriction fragment length polymorphism analysis of RNA transcripts using PCR (PCR–RFLP)

Reagents

5 mg/ml Bovine serum albumin (BSA)
100 mM dithiothreitol (DTT)
DNase I (RNase free; Promega)
dNTP stock (containing 10 mM dATP, dCTP, dGTP and dTTP)
Endonucleases (*Hae*III, *Msp*I, *Rsp*I and *Taq*I)
Ethanol⚠
Mineral oil
M-MLV reverse transcriptase (BRL)
0.5 M NaOH⚠
Oligo-dT (12–18) (Pharmacia)
Placental RNase inhibitor (USB)
Primer mix stocks at 10 µM

10× RT buffer (500 mM Tris-HCl (pH 8.3), 400 mM KCl, 60 mM MgCl$_2$)
SeaKem LE agarose
3 M Sodium acetate
10× Taq buffer (500 mM Tris-HCl (pH 8.3), 10 mM MgCl$_2$, 200 mM KCl)
Taq polymerase (5 units/µl)
Total RNA (approx. 5 µg)
T$_{10}$E$_{0.1}$ (10 mM Tris base, 0.1 mM EDTA, pH 7.5)

Equipment

Agarose gel electrophoresis equipment
Thermal cycler
Water baths at 37°C and 68°C

Technique (ref. 13)

1 Total RNA isolated using a commercially available kit (e.g. Glasmax RNA, Gibco) or using a standard protocol (*Protocol 1*) is used in the initial reverse transcription step. Add DNase I to a final concentration of 20 units/400 µg RNA and incubate at 37°C for 1 h.

2 Precipitate the RNA by adding 0.1 vol. 3 M sodium acetate and 2 vols ethanol at −70°C until needed[1]. Collect the RNA precipitate by centrifugation at 15 000 g for 15 min at 4°C. Wash the pellet with ice-cold 75% ethanol and resuspend in water at 1 µg total RNA/µl.

3 Reverse transcription: set up the reaction in a 0.5 ml microcentrifuge tube – 10 µl (10 µg) RNA, 5 µl 10× RT buffer, 5 µl DTT, 1 µl dNTP stock, 4 µl (45 units) placental RNase inhibitor, 4 µl (800 units) M-MLV reverse transcriptase, 2 µg oligo dT, 21 µl water. Incubate at 37°C for 1 h.

4 Incubate reactions in 0.5 M NaOH at 68°C for 20 min. Chloroform extract, ethanol precipitate and resuspend in 10 µl $T_{10}E_{0.1}$.

5 PCR reaction: 1 µl first strand reaction, 1 µl 10× Taq buffer, 2 µl dNTP stock, 1 µl primer mixture, 0.2 µl Taq polymerase and 4.8 µl water. Cover the reaction with 20 µl mineral oil. Heat denature the samples prior to PCR at 94°C for 2 min. Carry out 40 cycles: 92°C for 10 sec, 50°C for 20 sec, 72°C for 60 sec. Force complete extension by incubating at 72°C for 6 min after PCR has finished.[2]

6 Run PCR products in a 2% agarose gel: 1× TBE (*Protocol 14*), isolate the bands and recover the DNA using a kit (e.g. Geneclean, USB) or according to *Protocol 34*.

7 Reamplify using 2 ng DNA (1 µl), 1 µl 10× Taq buffer, 2 µl dNTP stock, 1 µl primer mixture, 0.2 µl Taq polymerase and 4.8 µl water. Cover the

Continued overleaf

Protocol 22. Restriction fragment length polymorphism

reaction with 20 μl mineral oil. Heat denature the samples prior to PCR at 94°C for 2 min. Carry out 30 cycles: 92°C for 10 sec, 55°C for 20 sec, 72°C for 60 sec. Force complete extension by incubating at 72°C for 6 min after PCR has finished.

8 Run 1 μl PCR products in a 2% agarose gel: 1× TBE (*Protocol 14*) and use the remainder in RFLP analysis.

9 Ethanol precipitate the PCR products[3] and wash with 75% ethanol. Resuspend in 20 μl $T_{10}E_{0.1}$ and digest 2 μl aliquots with each of the endonucleases according to procedures supplied with individual enzymes. Electrophorese the products on a 2% agarose gel: 1× TBE (*Protocol 14*) to define which enzyme produces a polymorphism in the samples you wish to compare.

Pause points

[1] May be left at −70°C for up to 24 h.
[2] May be kept at −20°C until needed.
[3] May be left at −20°C for up to 24 h.

Protocol 23. PCR ribotyping of bacterial rRNA genes – species/type identification

Reagents

Chloroform:isoamyl alcohol (IAA) (24:1)⚠

dNTP mix (20 mM dATP, dCTP, dGTP and dTTP)

Ethidium bromide (EtBr) solution⚠

0.5 M Ethylenediaminetetra-acetic acid (EDTA)

LB broth (10 g bacto-tryptone, 5 g bacto-yeast extract, 10 g NaCl/l (pH 7))

0.5 mg/ml Lysostaphin

4 mg/ml Lysozyme (freshly prepared)

Mineral oil

NuSieve 3:1 agarose–FMC

PCR primers (50 µM stocks)

1 mg/ml Proteinase K (fresh)

10% Sodium dodecyl sulfate (SDS)

Taq polymerase (5 units/µl)

Taq polymerase buffer (100 mM Tris-HCl, pH 8.3, 500 mM KCl, 30 mM MgCl$_2$)

10 mM Tris-HCl, pH 8.0, 5 mM EDTA

Equipment

Thermal cycler

Water baths at 37°C and 55°C

Technique (ref. 14)

1 Bacterial growth and lysis: grow bacteria overnight in LB broth and collect by centrifugation at 2000 *g* for 10 min at RT.

2 Resuspend the bacterial pellet in 1 ml 10 mM Tris-HCl buffer containing 5 mM EDTA. Collect the bacteria by centrifugation at 2000 *g* for 10 min at RT.

3 Resuspend the pellet in 300 µl 10 mM Tris-HCl, 5 mM EDTA. Add

Notes

① Primers should be complementary to conserved regions, e.g. 16S and 23S primers detailed in ref. 14.

50 μl lysozyme solution and 30 μl lysostaphin. Incubate at 37°C for 1 h.

4   Add 22 μl 10% SDS and 44 μl proteinase K. Incubate at 55°C until the solution clears.

5   Heat inactivate proteinase K at 95°C for 10 min. Phenol–chloroform and chloroform extract to remove the proteinase K. Calculate the DNA concentration spectrophotometrically (OD_{260} of 1.0 = 50 μg/ml DNA).

6   PCR reaction: 10 μl Taq buffer, 1 μl each primer①, 1 μl dNTP mix, 100 ng template DNA (step 3) and 0.5 μl Taq polymerase. Cover the reaction with 100 μl mineral oil. Amplify for 35 cycles (1 min at 94°C, 1 min at 50°C and 1 min at 72°C).

7   Recover the sample from the mineral oil by chloroform extraction. Analyze products on a 3% native agarose gel containing 5 μg/ml EtBr (*Protocol 14*).

Protocol 23. PCR ribotyping of bacterial rRNA genes

Protocol 24. Assessment of PCR fidelity using temperature gradient gel electrophoresis

Reagents

40% Acrylamide stock solution (37.5% acrylamide, 1% methylene bisacrylamide in water)⚠
10% (w/v) Ammonium persulfate (APS)
dNTP mix (20 mM dATP, dCTP, dGTP and dTTP)
Ethidium bromide (EtBr) stock (5 mg/ml)⚠
Fresh/deionized formamide⚠
Mineral oil
^{32}P-labeled PCR primer 1⚠
Plasmid DNA (template)
60 μM Primer 2
SeaKem agarose
Taq polymerase (5 units/μl)

10× Taq polymerase buffer (100 mM Tris-HCl, pH 8.3, 500 mM KCl, 25 mM MgCl$_2$)
10× TBE (0.9 M Tris base, 0.9 M boric acid, 20 mM ethylenediaminetetra-acetic acid (EDTA))
30 μl N,N,N',N'-tetramethylethylenediamine (TEMED)
Urea, molecular biology grade

Equipment

Nalgene filter (0.2-μm pore size)
Shielding
Temperature gradient gel electrophoresis (TGGE) equipment
Thermal cycler

Technique (ref. 15)

1 End label one of the primers with ^{32}P (*Protocol 5*).

2 PCR reaction: 10 μl Taq buffer, 1 μl (50 pg) plasmid DNA, 1 μl of each primer, 1 μl dNTP mix, 0.5 μl Taq polymerase, 84.5 μl water. Overlay with 100 μl mineral oil.

Notes

① This produces a denaturing 6.5% polyacrylamide gel (37.5:1 acrylamide:bisacrylamide), 4.2 M urea, 24% v/v formamide in 0.5× TBE.

3 Heat denature at 94°C for 3 min prior to PCR (94°C for 1 min, 44°C for 2 min and 72°C for 1 min for 30 cycles).

4 Assess product size and purity by electrophoresis on a 1% agarose gel (*Protocol 14*).

5 In a glass beaker, mix 6.15 ml 40% acrylamide stock solution, 12.6 g urea, 2.5 ml 10× TBE, 12 ml formamide and 10 ml sterile double-distilled water. Dissolve urea using a magnetic stirrer and make final volume to 50 ml with sterile double-distilled water. Filter the solution through a 0.2-μm pore Nalgene filter.①

6 Add 30 μl TEMED and 300 μl 10% APS, mix and quickly pour between the gel plates, avoiding bubble formation. Allow the gel to set at RT prior to mounting between the two aluminum heating blocks of the TGGE apparatus.

7 TGGE gels can be electrophoresed parallel to the temperature gradient or perpendicular to it (see figures on right). When carrying out parallel runs, samples are loaded into the 1-cm wells or during a perpendicular run the sample is loaded along a well spanning nearly the entire width of the gel.

8 Temperature gradients to be used vary with the sample DNA but gradients of 28°C to 32°C can be used as a starting point of optimization. Electrophoresis is carried out at 4.5–6 V/cm for a 20 cm gel for 14–18 h (overnight) in 0.5× TBE.

Continued overleaf

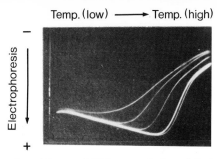

Reproduced from ref. 15 by permission of Oxford University Press.

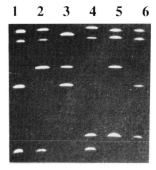

Reproduced from ref. 15 by permission of Oxford University Press.

Protocol 24. Temperature gradient gel electrophoresis

9 Staining: place the gel into electrophoresis buffer containing 5 µg/ml EtBr (1 µl EtBr/ml buffer). Fill a syringe with electrophoresis buffer and squirt between the gel and the plate to release the gel from the plate into a container. Leave the gel immersed at RT for 30 min.

10 Destain by carefully decanting EtBr solution off and replace with electrophoresis buffer for 5 min. Visualize the RNA bands by placing the gel on a UV transilluminator.

11 Record the image either using an image capture device (e.g. Mitsubishi video copy processor) or by Polaroid photography (typical settings: f4.5, 1/30–1/4 sec using Polaroid Polablue (135 mm) film).

Protocol 25. Pulsed field gel electrophoresis analysis of high molecular weight DNA

Reagents

Wash solution ($T_{10}E_1$; pH 7.6, 40 µg/ml phenylmethylsulfonyl fluoride (PMSF), see Appendix B)⚠ (freshly prepared)
Low-melt InCert agarose FMC
Lysis buffer (0.1 M ethylenediaminetetra-acetic acid (EDTA; pH 8), 10 mM Tris-HCl (pH 7.6), 0.02 M NaCl
Molecular weight marker plugs (e.g. lambda DNA ladders, FMC)
Phosphate-buffered saline (PBS; 0.2 M NaCl, 2.5 mM KCl, 10 mM Na_2HPO_4, 1.8 mM KH_2PO_4)
Proteinase K

20% Sarkosyl
SeaKem Gold agarose or equivalent
0.5× TBE (45 mM Tris, 45 mM boric acid, 1 mM EDTA, pH 8.3)

Equipment

CHEF-DR II Drive Module
Electrophoresis unit (Bio-Rad)
UV light box⚠
Water baths at 42°C and 50°C

Technique (ref. 16)

1 Lysis of cells containing DNA to be analyzed: wash cultured cells in ice-cold PBS three times and resuspend in lysis buffer at a concentration of 5×10^7 cells/ml in a final volume of 50 µl.①

2 Prepare an equal volume of 1% low-melting temperature agarose in lysis buffer and cool the melted agarose to 42°C.

3 Warm the cell suspension to 42°C and add the molten agarose to the

Notes

① Tissue samples (e.g. noncell culture mammalian, yeast and plant tissue) must be ground prior to resuspension.

② Makeshift Parafilm molds can be created on ice if preferable.

③ Some DNA molecules are too large to be separated by PFGE and are cut prior to electrophoresis with enzymes that cut only occasionally (e.g. *Not*I). The agarose insert is incubated in a restriction reaction prior to electrophoresis.

④ Agar blocks may be stored in 0.5 M EDTA (pH 8) at 4°C and can be washed in $T_{10}E_1$ prior to use.

cells and stir to mix. Pipette 100 μl of the mixture into a well mold (e.g. Plexiglas mold) and place at 4°C to set.②

4 Recover the agarose samples from the molds and transfer them to 5 ml lysis buffer supplemented with 1 mg/ml proteinase K and 1% sarkosyl. Incubate at 50°C for 48 h changing the lysis solution after 24 h. Transfer the blocks into 5 ml of freshly prepared wash solution for 2 h, changing the wash solution after 1 h.③

5 Make a 0.75% agarose gel (SeaKem Gold) with 0.5× TBE and allow to set at 12°C. Place the agarose sample and ladder blocks into the wells and seal with low-melt 0.5% agarose.④

6 Electrophorese at 12°C for 18 h in 0.5× TBE in a CHEF-DR II Drive Module at 200 V with a ramped pulse of 0.1 to 100 sec. To stain, immerse the gel in 5 μg/ml EtBr for 30 min.

7 Visualize the DNA bands by placing the gel on a UV light box and record the image either using an image capture device (e.g. Mitsubishi video copy processor) or by Polaroid photography (typical settings: f4.5, 1/30–1/4 sec using Polaroid Polablue (135 mm) film). The gel can be Southern blotted (*Protocol 26*) and probed (*Protocol 28*) as required.

Protocol 25. Pulsed field gel electrophoresis

VI DETECTION OF NUCLEIC ACID SPECIES – AUTORADIOGRAPHY AND FLUOROGRAPHY

Basic principles of autoradiography

Autoradiography may be direct or indirect. The latter is used for high energy isotopes such as ^{32}P that emit particles which mostly pass through X-ray film without exposing it. These emissions can be recorded on the film by placing an intensifying screen behind the film which absorbs the particles and produces light. This light is recorded on the film, superimposed over the low level exposure of the particles passing through the film. Isotopes such as ^{35}S do not penetrate the film and do not require intensifying screens and this is termed direct autoradiography. This involves the absorption of radiation by X-ray film, producing bands or foci at a position corresponding to the location of the radiation. Fluorography uses scintillants diffused into the gel to convert the weak emissions within the gel to light. Fluorography can be used to increase the sensitivity and detection of medium energy isotopes or can be used for detecting weak isotopes such as tritium which are not otherwise detectable by autoradiography.

Film types

The choice of film type is governed by laboratory budget and experimental needs. A film that has a colorless background (e.g. Hyperfilm) is a good choice for publishing data to obtain contrast between the signal and the background, particularly for low intensity signals. High intensity signals obtained from Southern blots or sequencing can be recorded using Kodak X-OMAT AR film, as background is not as important here. Preflashed X-ray film is used when quantitation is required as this improves the linearity of the exposure. Some films commonly used for autoradiography are listed in *Table 8*.

Apparatus (exposure and developing)

The method of assessment depends on the nature of the isotope being used (*Table 2*). Isotopes such as ^{32}P, ^{33}P and ^{125}I are strong enough emitters to penetrate the gel segment in which they are present. These isotopes emit particles of such strength

Table 8. Film types used in autoradiography

Film	Source[a]	Comment
AGFA	APP	
Fuji-AX	GRI	
GRI-AX	GRI	
Hyperfilm	AIP	Colorless film when developed
Hyperpaper	AIP	
Kodak X-OMAT AR	EKL, IBI, SCC	Commonly used for high abundance Southern blotting

[a] See Appendix C: Suppliers.

that few are absorbed by X-ray film but instead pass straight through the film, generating an image not representative of the amount of radiation present. Indirect autoradiography must be used to locate the radioactivity in gels containing these isotopes. This is routinely achieved by placing an 'intensifying' screen behind the X-ray film. Intensifying screens incorporate calcium tungstate which emits multiple flashes of light when particles impact on it. These produce photographic images overlaid over the autoradiographic image.

Filmless autoradiography is now possible using image analysis equipment. The Model GS–250 Molecular Imager System (Bio-Rad) enables imaging and quantitation of radioactive or chemiluminescent samples at a greater sensitivity than that attained by using film. A phosphor screen captures radioactivity and light which can be scanned to quantitate signals.

Weaker isotopes such as ^{3}H, ^{33}P, ^{35}S and ^{14}C cannot penetrate the gel matrix requiring the gel to be dried before autoradiographic assessment. Gels are usually dried by using commercial gel driers with heating elements. This type of apparatus can dry gels to a thin film in only 1–2 h. Gel driers that use vacuum systems only, can take up to 24 h to dry gels. Labeled species

Detection of nucleic acid species

containing ^{3}H in gels cannot be detected by autoradiography even if the gel is dried. Thus if a very weak isotope is being used (^{3}H, ^{14}C and ^{35}S) the emission can be enhanced using fluorography. This uses scintillants to convert the weak emissions to light that has an effect on the X-ray film.

Gel segments containing a strong emitter such as ^{32}P or ^{125}I can be measured in a scintillation counter without the aid of scintillants. This method of counting utilizes the strong emissions which cause Cerenkov radiation. This radiation is usually measured using the ^{3}H channel on a scintillation counter. The efficiency of counting is at most 50% and varies with gel segment size. Segment size distortions can be nullified using NaCl prior to counting (see ref. 11 in Chapter IV). Segments containing low level radiation need to be dried down on filter paper and measured against radioactive standards or the nucleic acid can be eluted, mixed with scintillants and counted in a scintillation counter.

Methods available

Blotting (see *Protocols 26* and *27*)
Blotting of gels allows nucleic acid samples separated by gel electrophoresis to be analyzed on a membrane using isotopic or nonisotopic probes. Gels can be blotted by capillary action where a solution is drawn through a gel carrying the nucleic acid with it, electroblotted by electrophoresing the nucleic acids from the gel on to the membrane or vacuum blotted where a vacuum forces the nucleic acid on to the membrane. The standard capillary blotting protocol is presented here

Problems: Uneven transfer of nucleic acid to the membrane, particularly when a gel has been loaded for equal amounts of nucleic acid,

References

1. Meinkoth, J. and Wahl, G. (1984) *Analyt. Biochem.* **138**:267.
2. Jones, P., Qiu, J. and Rickwood, D. (1994) *RNA Isolation and Analysis.* BIOS Scientific Publishers, Oxford.
3. Denhardt, D.T. (1966) *Biochem. Biophys. Res. Comm.* **23**:641.

Protocols provided

can ruin analysis. Gels should be run with ethidium bromide (even RNA gels) in to allow assessment of blotting on a UV transillumina-tor. Uneven transfer is usually due to uneven distribution of the weight on top of the blot (capillary blotting), inefficient transfer of high molecular weight DNA or inhomogeneous gels. Check transfer routinely to avoid artefacts.

Southern analysis (see *Protocol 28*)

There are two general procedures for Southern probing using either a simple hybridization solution or a complex hybridization solution. The simple solution is used in analysis involving large amounts of fixed target DNA (e.g. plasmid digests) and the complex solution is used for more demanding procedures (e.g. probing of genomic DNA digests). A discussion of the hybridization of nucleic acids to nucleic acids bound on solid supports can be found in ref. 1.

Problems: Radioactivity localized to only a portion of the membrane can be caused as a result of poor distribution of the probe over the membrane.

Northern analysis (see *Protocol 29*)

This technique is usually used to quantitate RNA, particularly for comparisons of tissues subject to different experimental regimes and tissues from different locations on an organism or from different organisms.

Detection of nucleic acid species

Problems: Quantitation must be carried out by reference to exposure standards as X-ray film exposes in a linear manner at first and then plateaus so a comparison of exposure of different bands can only be carried out for a certain range of band densities. This is essential for analyses involving equal loading of different samples.

Fluorography (see *Protocols 30a* and *b*)

Labeled species containing ^3H in gels cannot be detected by autoradiography even if the gel is dried. Thus if a very weak isotope is being used (^3H, ^{14}C and ^{35}S) the emission can be enhanced using fluorography. This uses scintillants to convert radioactive emissions to light such that they have an effect on the X-ray film. Some methods are outlined in *Protocol 30*.

Problems: Dimethylsulfoxide (DMSO) is toxic so carry out manipulations with care.

Autoradiography (see *Protocol 31*)

The length of film exposure depends on emission strengths. A ^{35}S sequencing gel, registering 50 d.p.s. (using a Geiger counter), will give an image after an overnight exposure, whilst a ^{33}P sequencing gel will take 5 h. If the radioisotope is already integral to the nucleic acid in a gel, autoradiographic assessment of location of the DNA/RNA in relation to size markers is possible.

Problems: If quantitation is the aim of an experiment then it must be carried out by reference to exposure standards as X-ray film exposes in a linear manner at first and then plateaus, so a comparison of exposure between different bands can only be carried out for a certain range of band densities.

Nonisotopic methods

Handling radioisotopes is hazardous. An alternative is to use a nonisotopically labeled probe as described in *Protocols 13a*, *b* and *c*.

Reagents

10× Saline sodium citrate (10 × SSC; 1.5 M NaCl, 0.15 M trisodium citrate, pH adjusted to 7.0)
Sterile double-distilled water

Equipment

Absorbent paper towels
Blotting tray
Glass support
Nylon/nitro-cellulose membrane
SaranWrap clingfilm
300 g Weight
Whatman 3MM filter paper

Technique (see figure on right)

1 Fill the blotting tray with 10× SSC (20× SSC for DNA).

2 Cut a section of membrane and five pieces of filter paper to the size of the gel. Cut a 1-cm triangle from the top right corner of each before immersing it in 2× SSC.①

3 Place a bridge support across the tray and lay a strip of filter paper ('wick'), equal in width to the gel, such that its ends enter the 10× SSC solution.

4 Pour 10× SSC across this filter wick to dampen it. Take the gel from the electrophoresis tank (*Protocol 14*) and cut a 1-cm triangle from the top right corner of the gel. If a denaturing agarose separation of RNA in a formaldehyde gel is being blotted then immerse the gel in sterile double-distilled water to wash excess formaldehyde away.

Notes

① Refer to manufacturer's instructions to see if pre-immersion is necessary as this varies with membrane type.

② This prevents any paper towels touching the wick.

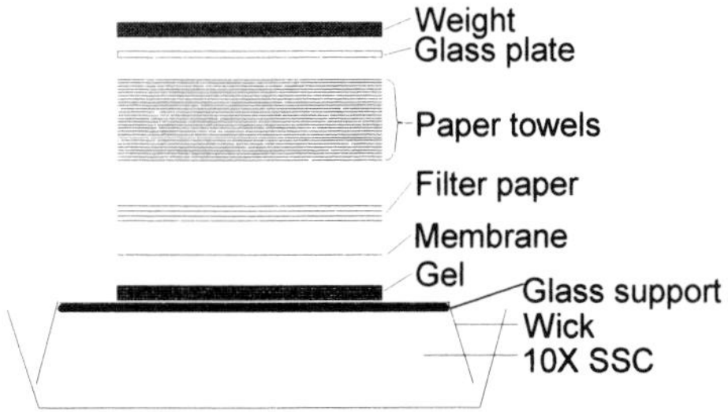

Weight
Glass plate
Paper towels
Filter paper
Membrane
Gel
Glass support
Wick
10X SSC

5 Place the gel face down on the filter paper wick. Lay the pre-soaked membrane over the gel, matching nicked orientation corners, excluding any bubbles below the membrane with gloved fingers.

6 Add five pre-soaked filter papers on to the membrane, excluding bubbles as before. Lay a piece of clingfilm across the entire tray and, using a razor blade, cut a central piece out of the film to expose only the gel/membrane/filter paper stack.②

7 Place absorbent paper towels on the stack to a height of about 4 cm. Lay a glass plate and 300-g weight on top of the towels and leave to capillary blot for 4–16 h.

8 Remove the membrane from the blot set-up and fix the RNA to the membrane by baking at 80°C for 2 h or by UV cross-linking (e.g. UV Stratalinker 1800 (Stratagene)) which cross-links the RNA with the membrane in 15–60 sec.

Protocol 26. Capillary blotting

Protocol 27. Electroblotting of polyacrylamide gels on to nylon membranes

Reagents

Electrophoresis buffer
Polyacrylamide slab gel to be blotted
6× Saline sodium citrate (6× SSC: 0.9 M NaCl, 90 mM trisodium
 citrate, pH 7.0)
Transfer buffer (0.1 M Tris-HCl (pH 7.8), 50 mM sodium acetate,
 5 mM ethylenediaminetetra-acetic acid (EDTA))

Equipment

Electroblotter (e.g. Transblot Cell, Bio-Rad)
Nylon membrane (e.g. Genescreen, NEN DuPont)
UV Stratalinker (e.g. Stratalinker 1800, Stratagene)
Whatman 3 mm filter paper

Technique

1 If the gel contained urea then immerse it in electrophoresis buffer for
 30 min to remove the urea.①

2 Place the gel in the transfer buffer for 10 min to equilibrate the gel.

3 Cut a piece of nylon membrane and Whatman 3MM filter paper to the
 size of the gel and soak in the transfer buffer. Nick one corner of the
 membrane for orientation purposes.

4 Fill the Transblot apparatus with the transfer buffer.

5 With the open Transblot cassette on the bench, place the filter paper on
 to one face followed by the gel and then the nylon membrane. Close and
 clamp the cassette and place into the Transblot apparatus.

Notes

① Wear gloves to prevent RNase contamination and transfer
 of DNA/RNA to the membrane.
② Keep blot damp during UV fixation.

6 Electroblot for at least 4 h at 200 mA at 4°C.

7 Recover the cassette and remove the membrane. Briefly rinse the membrane in 6× SSC and place RNA side up on to Whatman 3MM paper (pre-dampened with 6× SSC).

8 Fix RNA to the membrane using a UV Stratalinker. Store blot at 4°C in the dark until hybridization.②

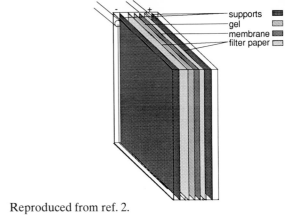

Reproduced from ref. 2.

Protocol 28. Southern probing for low and high abundance target sequences

Reagents

Complex hybridization solution (50% formamide, 6× SSPE, 5× Denhardt's, 1% SDS)⚠

100× Denhardt's, 500 ml (10 g Ficoll, 10 g polyvinylpyrrolidone, 10 g bovine serum albumin (BSA), filter sterilized)

20× Saline sodium citrate (20× SSC: 3 M NaCl, 300 mM trisodium citrate, pH 7.0)

Simple hybridization solution (1 M NaCl, 1% sodium dodecyl-sulfate; SDS)

20× SSPE (3 M NaCl, 0.177 M NaH_2PO_4, 0.02 M ethylene-diaminetetra-acetic acid (EDTA), pH 7.4)

Wash solution 1: 1× SSC, 1% SDS

Wash solution 2: 0.1× SSC, 0.1% SDS

Equipment

Hybridization vessel

Orbital 65°C incubator

Water bath at 65°C

Boiling water bath

Technique (ref. 3)

1 Place the Southern blot membrane into a hybridization chamber/bottle/bag with 1 ml of the appropriate hybridization solution/10 cm^2 of membrane. Seal the hybridization vessel and place at 65°C for 1 h ('pre-hybridization') with constant movement of the solution over the membrane (e.g. 20 r.p.m.). Prewarm excess hybridization solution (e.g. 50 ml) to 65°C in a polypropylene tube.①

2 Heat denature the radiolabeled probe (*Protocols 11/12*) in a screw-capped tube at 95°C for 5 min and snap cool on ice. Discard the pre-hybridization solution and replace with 0.5 ml of hybridization solution/

Notes

① Herring sperm DNA is given here as the blocking agent. This can be replaced with many other DNA sources such as salmon sperm DNA and calf thymus DNA.

② If plastic bags are used bubbles should be excluded before sealing.

③ Wash conditions may be made less stringent by lowering wash 2 temperature.

10 cm^2 of membrane. Pipette the probe into the hybridization mixture avoiding touching the membrane with the tip.②

3 The hybridization chamber/bottle/bag is sealed and placed at 65°C overnight ('hybridization') with constant circulation of solution over the membrane.

4 Remove the membrane from the vessel and immerse in wash solution 1. Shake gently on an orbital table for 10 min at RT. Discard the hybridization solution according to regulations. Prewarm wash solution 2 to 65°C.③

5 Repeat the wash in wash solution 1 and then in wash solution 2 at 65°C. Recover the membrane and drain off excess liquid prior to autoradiography (*Protocol 31*).

Protocol 28. Southern probing

Reagents

100× Denhardt's, 500 ml (10 g Ficoll, 10 g polyvinylpyrrolidine,
  10 g bovine serum albumin (BSA), filter sterilized)
Hybridization solution (50% formamide, 6× SSPE, 5× Denhardt's,
  1% sodium dodecyl sulfate (SDS), 0.1 mg/ml (sonicated)
  Herring sperm DNA)⚠
20× Saline sodium citrate (SSC) (3 M NaCl, 300 mM trisodium
  citrate, pH 7.0)
20× SSPE (3 M NaCl, 0.177 M NaH_2PO_4, 0.02 M ethylenediaminetetra-acetic acid (EDTA) pH 7.4)

Wash solution 1: 1× SSC, 1% SDS
Wash solution 2: 0.1× SSC, 0.1% SDS

Equipment

Boiling water bath
Hybridization vessel
Orbital 42°C incubator
Water bath at 42°C

Technique (ref. 3)

1 Place the northern blot membrane into a hybridization chamber/bottle/
   bag with 1 ml of hybridization solution/10 cm^2 of membrane. Seal the
   hybridization vessel and place at 42°C for 1 h ('prehybridization') with
   constant movement of the solution over the membrane (e.g. 20 r.p.m.).
   Prewarm the excess hybridization solution (e.g. 50 ml) to 42°C in a
   polypropylene tube.①

2 Heat denature the radiolabeled probe (*Protocols 11/12*) in a screw-
   capped tube at 95°C for 5 min and snap cool on ice. Discard the

Notes

① Herring sperm DNA is given here as the blocking agent.
   This can be replaced with many other DNA sources such
   as salmon sperm DNA and calf thymus DNA.
② If plastic bags are used bubbles should be excluded before
   sealing.
③ Wash conditions may be made less stringent by lowering
   wash 2 temperature.

prehybridization solution and replace with 0.5 ml of hybridization solution/10 cm^2 of membrane. Pipette the probe into the hybridization mixture avoiding touching the membrane with the tip.

3 The hybridization chamber/bottle/bag is sealed and placed at 42°C overnight ('hybridization') with constant circulation of the solution over the membrane.②

4 Remove the membrane from the vessel and immerse in wash solution 1. Shake gently on an orbital table for 10 min at RT. Discard the hybridization solution according to regulations. Prewarm wash solution 2 to 65°C.③

5 Repeat the wash in wash solution 1 and then in wash solution 2 at 42°C. Recover the membrane and drain off excess liquid prior to autoradiography (*Protocol 31*).

Protocol 29. Northern probing

Protocol 30a: Fluorography using PPO with polyacrylamide gels

Reagents

Dimethyl sulfoxide (DMSO)⚠
20% PPO ((w/w) dissolved in DMSO)⚠

Equipment

Gel drier
Gel trays
Preflashed Kodak X-OMAT R film

Technique

1 Immerse the stained or unstained polyacrylamide gel in 300 ml DMSO for 30 min, decant and repeat.①

2 Immerse the gel in 4× gel vols of 20% PPO for 3 h.

3 Wash the gel in 20 vols of water for at least 1 h to remove DMSO and precipitate PPO in the gel.

4 Dry the gel under vacuum.

5 Place the dried gel in contact with preflashed Kodak X-OMAT R film at −70°C.

6 Develop the film after the defined amount of time according to the instructions supplied with developing chemicals/apparatus.

Notes

① This is the method of choice for ^{14}C and ^{35}S.

Protocol 30b. **Fluorography using sodium salicylate**

Reagents

1.0 M Sodium salicylate (pH 7.0)
5% Trichloroacetic acid (TCA) or methanol:acetic acid:water
  (5:1:5 by vol.)⚠

Equipment

Gel drier
Gel trays
Preflashed Kodak X-OMAT R film
Whatman 3MM paper

Technique

1 Fix the gel if required, in 5% TCA or methanol:acetic acid:water (5:1:5 by vol).①

2 If the gel has been fixed, the acid must be removed by immersing the gel in 20 vols of distilled water for 30 min; to prevent sodium salicylate precipitation.

3 Immerse the gel in 10 vols of 1.0 M sodium salicylate (pH 7.0) for 20 min.

4 Lay the gel on moist Whatman 3MM paper and dry under vacuum.

5 Place the dried gel in contact with preflashed Kodak X-OMAT R film at −70°C.

6 Develop the film after the defined amount of time according to the instructions supplied with developing chemicals/apparatus.

Notes

① This is the method of choice for ^{3}H and gel segments can be rehydrated in electrophoresis buffer allowing second dimension analysis.

Protocol 31. Detection of radioactivity using autoradiography

Reagents

Radioactive membrane/gel⚠
X-ray film developing chemicals

Equipment

Autoradiographic cassette
Clingfilm (e.g. SaranWrap)
Dark room and safe-light
Intensifying screen
Optional automatic developer
X-ray film

Technique

1 Place the membrane/gel into the cassette and cover with a piece of clingfilm to prevent the film becoming wet.
2 Turn the room lights off and the safe-light on.
3 Remove X-ray film from the box and secure the box shut with tape before placing the film into the cassette in order to avoid accidental exposure of the entire stock of film.①
4 Cut a corner from the film and place into the cassette such that the cut corner lies in the top right. This allows orientation after developing.
5 Cover the film with intensifying screen and close the cassette.
6 Store the cassette at $-70°C$ if high emission strength sources are being used (^{32}P) or RT for isotopes such as ^{35}S.②
7 Develop the film after the defined amount of time according to the instructions supplied with developing chemicals/apparatus.

Notes

① X-ray film is usually supplied in a light-tight envelope within a box. It is good practice to slit one end of the envelope open, storing this open end at the bottom of the box.
② High emission sources placed at room temperature produce ill-defined images which are represented as tight bands if stored at $-70°C$.

VII ISOLATION OF NUCLEIC ACIDS FROM GELS

Gel electrophoresis is often an intermediate step in many protocols which require the nucleic acid to be recovered from the gel for further analysis or manipulation. The methods used to recover nucleic acids from gels rarely yield more than 80% of the nucleic acid in a defined gel segment. This must be adjusted for during experimental design, to ensure that enough nucleic acid is carried through each step of the protocol.

Methods available

Simple diffusion (see *Protocols 32* and *33*)
Many methods of elution by diffusion are available for both RNA and DNA. Elution is best achieved with nucleic acid of less than 5 kb in length. Larger species are recovered at efficiencies of less than 50%. Elution is also best achieved if large amounts of DNA/RNA are loaded into the gel due to the loss during recovery.
Problems: Many methods are labor intensive which limits the number of samples that can be processed at the same time.

Electroelution (see *Protocol 34b*)
Electroelution uses the migration of charged molecules in much the same way as gel electrophoresis to transfer nucleic acid from a gel segment into the buffer around the segment. The nucleic acid is contained in a semi-permeable dialysis tubing which allows recovery after elution.

References

1. Wieslander, J. (1979) *Analyt. Biochem.* **98**:305.
2. Dolja, V.V., Negruk,V.I., Novikov, V.K. and Atabekov, J.C. (1977) *Analyt. Biochem.* **80**:502.

Protocols provided

32. *Recovery of high molecular weight RNA from agarose gels*
33a+b. *Recovery of RNA from polyacrylamide gels*
34a. *Recovery of DNA from low melting temperature agarose*
34b. *Recovery of DNA from polyacrylamide or agarose by electroelution*
35. *Recovery of DNA from polyacrylamide gels: crush and soak method*

Problems: Nucleic acid can adhere to the walls of the dialysis tubing, reducing recovery. The reversal of current at the end of the procedure limits this but checking on a UV transilluminator prior to removal of the buffer is recommended.

Low melting temperature agarose (see *Protocol 34a*)

The use of low melting temperature agarose allows easier recovery of nucleic acid and allows some enzyme reactions to be carried out in the melted gel.

Problems: This method is not as reproducible as electroelution and can be troublesome if high currents are used through the gel, leading to premature melting during electrophoresis.

Protocol 32. Recovery of high molecular weight RNA from agarose gels

Reagents

Buffer-saturated phenol⚠, molecular biology grade
Electrophoresis buffer
Ethanol⚠
1 M NaCl

Equipment

Microcentrifuge
Water bath at 65°C

Technique

1 Place the gel slices into a tube with 5–10 vol. of electrophoresis buffer and warm to 65°C for 5 min to melt the agarose.①
2 Extract with an equal volume of buffer-saturated phenol for 15 min and centrifuge at 14 000 g for 10 min. Recover the aqueous layer.
3 Re-extract with an equal volume of buffer-saturated phenol for 5 min and centrifuge at 14 000 g for 10 min. Recover the aqueous layer.
4 Add 0.1 vol. of 1 M NaCl and 2.5 vol. ethanol and place at –20°C for at least 2 h. Collect the RNA precipitate by centrifugation at 14 000 g for 10 min at RT.

Notes

① This method relies on the use of low melting point agarose.

Protocol 33a. Recovery of RNA from polyacrylamide gels

Reagents

Autoclaved double-distilled water
1 M CaCl$_2$
Ethanol⚠
70% Ethanol⚠
20 mM Ethylenediaminetetra-acetic acid (EDTA) pH 8.0
20 mM Glycine (pH 9), 0.05% sodium dodecyl sulfate (SDS)⚠

4 M KCl
20 mM Tris-HCl (pH 7.8)

Equipment

Benchtop centrifuge
Centrifuge for 23 000 g spins

Technique (ref. 2)

1 Freeze the gel segment for 10 min and thaw at RT. Homogenize the gel segment in 0.5 ml 20 mM glycine (pH 9), 0.05% SDS and mix overnight at 4°C. Centrifuge at 23 000 g for 15 min at 4°C and recover the supernatant.①

2 Repeat the extraction twice with 0.25 ml 20 mM glycine (pH 9), 0.05% SDS and pool the supernatants.

3 Add 60 μl 4 M KCl to precipitate the SDS, centrifuge at 14 000 g for 10 min at 4°C and recover the supernatant. Adjust pH of the supernatant to 5.5–6.0.

4 Add 1/5 vol. 1 M CaCl$_2$ to precipitate the RNA and leave at 0°C

Notes

① This method can be used to obtain high purity RNA but may have a low recovery. Use only when RNA is plentiful or high purity is required.

② Redissolving in EDTA and precipitation produces RNA free of polyacrylates, SDS and other gel contaminants not removed by normal ethanol precipitations.

overnight. Collect the RNA by centrifugation at 23 000 *g* for 15 min at 4°C. Dissolve the pellet in 200 μl 20 mM EDTA.② ⒈

5 Add 200 μl 20 mM Tris-HCl (pH 7.8) and precipitate the RNA with 2 vols (800 μl) ethanol at –20°C for 2 h. Repeat steps 4 and 5 once.

6 Wash the final pellet in 70% ethanol and air dry. Resuspend in the required volume of autoclaved double-distilled water.

Pause point

⒈ May be left at –20°C for up to 24 h.

Protocol 33b.

Reagents

Buffer-saturated phenol:chloroform:isoamyl alcohol
  (IAA)(25:24:1)⚠
Ethanol⚠
3 M Sodium acetate (pH 5.6)
0.3 M Sodium acetate (pH 6.0), 0.2% sodium dodecyl sulfate
  (SDS)⚠
Sterile double-distilled water

Equipment

Microcentrifuge
Rotary shaker
Sterile siliconized glass rod
Whirlimixer

Technique

1 Cut the gel segment from gel and place into a 15 ml glass tube. Crush the segment with a sterile siliconized glass rod.

Continued overleaf

Notes

① This protocol is quicker than *Protocol 33a* but may not yield high purity RNA. The loss of RNA, however, should be less than that obtained in *Protocol 33a*.

Protocols 33a+b. Recovery of RNA from polyacrylamide gels

2 Add 0.4 ml 0.3 M sodium acetate (pH 6.0), 0.2% SDS and shake for at least 3 h on a rotary shaker.

3 Collect the polyacrylamide debris by centrifugation at 14 000 g for 2 min.

4 Recover the aqueous layer to a fresh tube and add 400 µl of buffer-saturated phenol:chloroform:IAA. Mix using a whirlimixer and centrifuge at 14 000 g for 2 min.

5 Recover the aqueous phase and precipitate the RNA by adding 1/10 vol. 3 M sodium acetate (pH 5.6) and 2.5 vols ethanol. Place at –20°C for at least 3 h and collect by centrifugation at 14 000 g for 10 min. [1]

6 Redissolve in the required volume of sterile water.

② Small quantities of RNA can be more effectively recovered by ultracentrifugation at 60 000 g for 10 min at 4°C.

Pause point

[1] May be left at –20°C for up to 24 h.

Protocol 34a. Recovery of DNA from low melting temperature agarose

Reagents

Buffered phenol⚠
Chloroform:isoamyl alcohol(IAA) (24:1)⚠
Ethanol⚠
200 mM NaCl
Sodium acetate (pH 5.6)
TE (pH 7.5; 10 mM Tris-HCl, 1 mM ethylenediaminetetra-acetic acid (EDTA))

Equipment

Glass rod
Microcentrifuge
Water bath at 65–70°C

Technique

1 Isolate the band of interest from agarose gel by cutting the band from the gel. Place the gel in a microcentrifuge tube containing an equal volume of 200 mM NaCl.①

2 Gently macerate the gel segment and place at 65–70°C, shaking occasionally until the gel has melted.

3 Phenol extract with an equal volume of phenol, shaking for 20 min prior to separation of phases by centrifugation at 15 000 g for 10 min. Recover the upper aqueous layer avoiding agarose precipitate at the phase interface. Repeat phenol extraction.

Notes

① Assume 1 g of agarose is equivalent to 1 ml.

4 Remove traces of phenol with a chloroform extraction. Add an equal volume of chloroform:IAA, shake and separate phases by centrifugation at 15 000 *g* for 5 min.

5 Precipitate the DNA by adding 1/10 vol. sodium acetate and 2.5 vols ethanol. Place at –20°C for at least 1 h ☐1. Collect DNA precipitate by centrifugation at 14 000 *g* for 10 min. Resuspend in required volume of TE.

Pause point

☐1 May be left at –20°C for up to 24 h.

Protocol 34b. Recovery of DNA from polacrylamide or agarose by electroelution

Reagents

Buffered phenol⚠
Chloroform:isoamyl alcohol (IAA) (24:1)⚠
Ethanol⚠
Sodium acetate (pH 5.6)
1× TBE electrophoresis buffer (90 mM Tris, 90 mM Boric acid, 2.5 mM ethylenediaminetetra-acetic acid (EDTA))
TE (10 mM Tris-HCl, 1 mM EDTA, pH 7.6)

Equipment

Dialysis tubing
Dialysis tubing clips
Electrophoresis tank
UV transilluminator⚠

Technique

1 Isolate the band of interest from the agarose gel by cutting the band from the gel and place into a piece of dialysis tubing. Clip one end of the tubing close to the gel.①

Continued overleaf

Notes

① Do not use this method on gels containing a phosphate-containing buffer.

2 Place 0.5–1 gel volume of electrophoresis buffer into the tubing. Remove bubbles by manipulating with gloved hands. Clip the tubing closed, close to the gel slice and orientate the slice to one side of the tubing.②

3 Place into an electrophoresis tank containing electrophoresis buffer with the gel slice closest and parallel to the cathode (see figure on right).

4 Electrophorese at 80 V for 30 min. Reverse the electrodes and electrophorese at 80 V for 30 sec to release DNA from the far wall of the tubing. Visualize the DNA on a UV transilluminator to ensure complete elution.

5 Recover the liquid and phenol extract, chloroform extract and ethanol precipitate.⬛1⬛ Resuspend in required volume of TE.

② Prepare dialysis tubing by boiling in 2% NaHCO$_3$, 1 mM EDTA for 10 min, rinse in water and boil in water for 10 min. Store in 50% ethanol at 4°C.

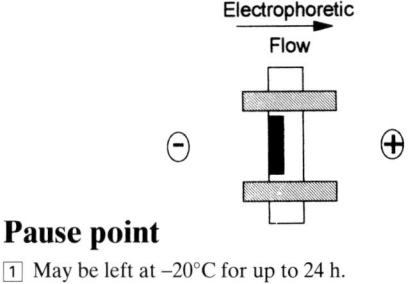

Pause point

⬛1⬛ May be left at −20°C for up to 24 h.

Protocol 35. Recovery of DNA from polyacrylamide gels: crush and soak method

Reagents

Elution solution (0.5 M ammonium acetate, 10 mM magnesium acetate, 1 mM ethylenediaminetetra-acetic acid (EDTA; pH 8), 0.1% sodium dodecyl sulfate (SDS))⚠

Ethanol⚠

3 M Sodium acetate (pH 5.2)

TE (10 mM Tris-HCl, 1 mM EDTA) pH 7.6

Equipment

Microcentrifuge

37°C oven with rotary table

Whatman GF/C filter

Technique

1 Isolate the band of interest from the agarose gel by cutting the band from the gel and place into a microcentrifuge tube. Use a Gilson pipette tip to crush the segment. Weigh the slice and tube and calculate the gel volume.①

2 Add 1–2 vols of elution solution and incubate at 37°C on a rotary table for 3–16 h. Small fragments (<500 bp) elute in 3–4 h and larger fragments in 12–16 h.

3 Centrifuge at 15 000 g for 1 min at 4°C, recover the supernatant and keep on ice. Add 0.5 vol. of elution solution to the polyacrylamide pellet,

Notes

① Assume 1 g of agarose gel is equivalent to 1 ml.

② To improve recovery of DNA add 10 µg of carrier RNA to the precipitation reaction.

vortex and centrifuge as before. Recover the supernatant and join with the supernatant on ice.

4 Pass solution through a disposable column containing a Whatman GF/C filter.

5 Add 2 vols of ethanol to precipitate the DNA and place on ice for 30 min. Recover the precipitate by centrifugation at 15 000 g for 10 min at 4°C. Redissolve in 200 µl TE (pH 7.6) and add 20 µl sodium acetate and 400 µl ethanol. Place on ice for 30 min and recover the precipitate as before.② 1

6 Wash the pellet with ice-cold 70% ethanol and redissolve the pellet in 10 µl TE.

Pause point

1 May be left at −20°C for up to 24 h.

Protocol 35. Recovery of DNA from polyacrylamide gels

VIII ANALYSIS OF NUCLEIC ACID:PROTEIN INTERACTIONS

The interaction of nucleic acids and proteins is essential to many molecular processes, ranging from gene transcription, DNA and RNA stability to enzymic catalysis. Analysis of the interface of the interaction(s) often is used as a first step in the investigation of the nucleic acid:protein complex. Analyses such as gel retardation and DNA footprinting allow the definition of a complex and can locate the interaction to a defined area of the RNA/DNA sequence.

Methods available

Cross-linking of nucleic acids and proteins
(see *Protocols 36a* and *b*)
Cross-linking of proteins to nucleic acids allows the observation of interactions of direct contacts between proteins: nucleic acids or allows the observation of potential interaction between neighboring proteins and ligands. Cross-linking procedures should be irreversible by conditions that occur during experimental analysis. The main methods of achieving this are by UV cross-linking or chemical cross-linking. UV cross-linking [1] is more controllable than chemical cross-linking and not as hazardous. One chemical cross-linking procedure, using formaldehyde, is, however, reversible which allows recovery of components after analysis [2, 3].

Gel retardation (see *Protocols 38* and *39*)
Basic principles: A solution containing protein–nucleic acid com-

References

1. Steige, W., Glotz, C. and Brimacombe, R. (1983) *Nucleic Acids Res.* **11**:1678.
2. Jackson, W. (1978) *Cell*, **15**:945.
3. Ip, Y.T., Jackson, V., Meier, J. and Chalkley, R. (1988) *J. Biol. Chem.* **263**:14 044.
4. Jarrell, K.A., Dietrich, R.C. and Perlman, P.S. (1988) *Mol. Cell. Biol.* **8**:2361.
5. Yonezawa, A. and Sugiura, Y. (1994) *Biochim. Biophys. Acta*, **1219**:369.
6. Latham, K.A., Dodson, M.L. and Lloyd, R.S. (1994) *J. Cellular Biochem.* S18c sic: 148 (meeting abstract).
7. Brimacombe, R., Greuer, B., Gulle, H., Kosack, M., Mitchell, P., Osswald, M., Stade, K. and Stiege, W. (1990) In *Ribosomes and Protein Synthesis – A Practical Approach* (G. Spedding, ed.), pp. 130–160. IRL Press, Oxford.
8. Kass, S., Tyc, K., Steitz, J.A. and Sollner-Webb, B. (1990) *Cell*, **60**:897.
9. Miller, K. and Sollner-Webb, B. (1981) *Cell*, **27**:165.

plexes and unbound molecules can be separated by electrophoresis on the basis of size. Free radiolabeled DNA molecules will migrate rapidly through the gel and constitute the 'free' band. The complexes will migrate more slowly through the gel and thus the name 'gel-retardation' or 'band shift assay'.

Binding conditions: Complex formation is usually in flux with dissociation and association continually occurring. A mixture of protein will bind to almost any DNA making it important to define conditions under which only specific interaction will occur. These conditions are usually defined by salt concentrations. Nucleic acid binding proteins are usually only partially active in that only 5–75% of the binding protein is active. Therefore an excess of protein is included in the assay to make up for this shortfall. DNA and RNA preparations, usually from plasmids, are commonly used in this type of assay and are totally available for binding. Complex formation will occur between virtually any protein and a piece of DNA/RNA if stringent conditions are not applied to binding. This is accomplished by adding a large excess of a nonspecific molecule such as poly-dI-dC (DNA shifts) or tRNA (RNA shifts).

Electrophoresis and analysis: Complexes are larger than free nucleic acid making it necessary to use less concentrated gels. Polyacrylamide systems are usually used at a gel concentration of about 5%. The complex is formed in solution prior to electrophoresis and often warmed to aid complex formation.

Protocols provided

36a. *Cross-linking of nucleic acids and proteins – UV cross-linking of ribosomes and RNA*

36b. *Cross-linking of nucleic acids and proteins – using formaldehyde*

37. *Antibody affinity isolation of nucleoprotein complexes*

38. *Gel retardation of rRNA:RNA processing complex*

39. *Gel retardation of DNA:protein complexes*

40. *DNA footprinting using DNase I*

41. *Electrophoresis of ribosomes – separation in one dimension*

42. *Electrophoresis of ribosomes and nucleosomes – separation in two dimensions*

43. *Electrophoresis of spliceosomes using composite gels*

Isolation of nucleic acid:protein interactions

Competitor analysis of nucleic acid: protein interactions: Competitor analysis is often used to define a specific interaction. In this system nonlabeled nucleic acid of exactly the same sequence as that used in the complex formation is used as the competitor. Increasing amounts of this competitor are added to separate reactions at 10–100 molar excess. A specific competitor will displace the radiolabeled probe indicating a specific interaction, whereas a nonspecific competitor will not displace the probe nucleic acid.

Most of the catalytic reactions involving RNA involve a protein component which can be analyzed in conjunction with the RNA using gel shift analyses. Competitor studies using nonspecific and specific RNA molecules allow the definition of an interaction [4] in much the same way as in DNA:protein gel shift assays.

Problems: Nuclease can often lead to shifted bands not being present as the nucleic acid component has been degraded. This can be avoided by inclusion of a nuclease inhibitor or fractionating the nuclear protein to obtain binding protein and no nucleases prior to binding.

DNA footprinting (see *Protocol 40*)
DNA footprinting is used as an in-depth focus upon the specific bases of DNA with which a binding protein interacts. This method is not only used to define the area of DNA that the protein binds to but may allow the determination of which face of the helix is interacting with the various binding proteins [5]. The DNA binding proteins that are

usually analyzed are ones which interact with promoter sequences but restriction nucleases have also been analyzed using this technology [6].

Problems: The definition of a specific complex can be complicated by nonspecific binding if binding conditions are not stringent. Increasing stringency of binding to its maximum helps to define a specific interaction. Binding may require other components such as ATP.

Electrophoresis of nucleoproteins (see *Protocols 41–43*)

The gel electrophoretic separation of bacterial polyribosomes (polysomes) and ribosomes is made possible by the development of composite gels containing both agarose and polyacrylamide. The addition of 0.5% (w/v) agarose to low-percentage polyacrylamide gels forms a mechanically stable, yet very porous, gel. Gels composed of 3.0% (w/v) polyacrylamide and 0.5% (w/v) agarose can be handled easily, and gels with even lower polyacrylamide concentrations can be used if necessary. The high resolution provided by these gels permits the separation of large macromolecules which differ only in conformation or by the presence of a single ribosomal protein. The separation is on the basis of molecular sieving and the extent of separation if determined primarily by the structure (size and shape) of the particle rather than by its charge.

The structure of ribosomes and polysomes depends on the surrounding ionic conditions, particularly the Mg^{2+} concentration. The inert matrix of agarose–polyacrylamide composite gels is particularly well-suited for

Isolation of nucleic acid:protein interactions

permitting variations in the ionic conditions so as to permit the study of the ribosome structure. As an example, polysomes from bacteria, which migrate intact in a gel prepared in a buffer containing 1–10 mM magnesium ions, will dissociate and migrate as 30S and 50S ribosomal subunits if the magnesium ion concentration is reduced to 0.2 mM or less.

The numerous methods available for preparing ribosomes and polysomes have been described in ref. 6. For optimal resolution of polysomes, the sample should be mixed with an equal volume of warm (50°C) 0.5% (w/v) agarose in buffer and allowed to set in a sample well from which the electrophoresis buffer has been removed. Samples not loaded as a gel can segment in a well and give a streaked but otherwise identical pattern. Gelling of the sample in the well is not necessary for the analysis of ribosomal subunits in gels containing a low concentration (0.2 mM) of magnesium ions as streaking does not occur for these small ribonucleoproteins. In these cases, the addition of sucrose to a final concentration of 10% (w/v) is sufficient for layering the samples into the wells after the reservoir buffer has been added to the electrophoresis apparatus. Bromophenol blue (0.1% (w/v)) can be added to the samples as a dye marker as it aids in visualizing the samples during application.

Problems: Dissociation of complexes can occur if ionic condition of the electrophoresis buffer interferes with binding. Optimize electrophoresis conditions to avoid this.

Protocol 36a. **Cross-linking of nucleic acids and proteins – UV cross-linking of ribosomes and RNA**

Reagents

Dilution buffer (50 mM KCl, 5 mM magnesium acetate, 25 mM triethanolamine-HCl (pH 7.8))
Ethanol⚠
1 M NaCl
Protease buffer (10 mM Tris-HCl (pH 7.8), 0.1% sodium dodecyl sulfate (SDS), 1 mM ethylenediaminetetra-acetic acid (EDTA))
10 mg/ml Proteinase K (Merck)
1 M Sodium acetate
10 mg/ml tRNA
Water-saturated phenol⚠

Equipment

Rotary table in cold room
Small petri dish
Swing-out centrifuge
UV lamps to deliver 25 $J/m^2/sec$
Water bath at 37°C

Technique (ref. 1)

1 Dilute the complex in the dilution buffer to a final concentration of 5 A_{260}/ml, and place in a small petri dish to a depth of 1 cm (light path depth). Irradiate from above with UV lamps at 4–6 cm from the sample for 2–10 min at 25 $J/m^2/sec$.

2 Precipitate the ribosomal subunits by adding 1/10 vol. 1 M sodium acetate and 2 vols ethanol at –20°C for 2 h. Collect by centrifugation at 10 000 *g* in a swing-out rotor for 30 min. Slowly decant off the

Continued overleaf

Note

① Adapt this protocol for optimum binding conditions defined in your experiments and precipitate your samples as defined by your system.

supernatant and wash the pellet with ice cold 80% ethanol. Allow the pellet to air dry and resuspend in 200 μl protease buffer. [1]

3 Add 200 μl 1 M NaCl and 40 μl proteinase K. Incubate at 37°C for 30 min. Add another 40 μl proteinase K and repeat the incubation. Add 2 μl tRNA and 500 μl phenol and shake at 4°C for 45 min on a rotary table.

4 Centrifuge at 5000 *g* for 10 min in a swing-out rotor and recover the aqueous phase. Ethanol precipitate once according to step 2. Redissolve in a solution appropriate to further experimentation.

Pause point

[1] May be left at −20°C for up to 24 h.

Protocol 36b. Cross-linking of nucleic acids and proteins – using formaldehyde

Reagents

DNA/RNA:protein complex
37% Formaldehyde⚠
Proteinase K (0.5 mg/ml)
10% Sodium dodecyl sulfate (SDS)⚠

TE (10 mM Tris base, 1 mM ethylenediaminetetra-acetic acid (EDTA) pH 7.5)

Equipment

Dialysis tubing
Water bath at 100°C/37°C

Technique (refs 2 and 3)

1 Form the DNA/RNA:protein complex according to your defined protocol and add 1/37 vol. formaldehyde. Place on ice for 24 h (in a cold room).

2 Remove the formaldehyde by dialysis against TE for 12 h. Ethanol precipitate and carry out experimental analysis as desired.[1]

3 Reversal of cross-linking by heating: heat at 100°C for 60 min or by proteolysis with 1/10 vol. 10% SDS and 1/10 vol. proteinase K at 37°C for 30 min.

4 If proteolysis was chosen then phenol extract and ethanol precipitate as detailed in *Protocol 34a* (steps 3 and 4).

Pause point

[1] May be left at –20°C for up to 24 h.

Protocol 37. Antibody affinity isolation of nucleoprotein complexes

Reagents

Affinity buffer (20 mM Tris-HCl (pH 7.0), 0.9% NaCl, 0.05% Tween 20)
Anti-nucleoprotein antisera
Complex buffer (20 mM Tris-HCl (pH 7.8), 0.1% sodium dodecyl sulfate (SDS), 6 mM 2-mercaptoethanol)
Goat anti-IgG immobilized on agarose
Binding buffer (20 mM Tris-HCl (pH 7.8), 100 mM NaCl, 0.05% Tween 20)
1 M and 3 M NaCl
^{32}P-labeled nucleoprotein complex⚠

Protease buffer (10 mM Tris-HCl (pH 7.8), 0.1% SDS, 1 mM ethylenediaminetetra-acetic acid (EDTA))
10 mg/ml Proteinase K (Merck)
1% SDS
10 mg/ml tRNA
Water-saturated phenol⚠

Equipment

Rotary end-over-end mixer
Scintillation counter
Swing-out benchtop centrifuge

Technique (ref. 7)

1 Shake the goat anti-IgG:agarose to form a slurry and transfer aliquots to a series of 0.5-ml microcentrifuge tubes. Allow the agarose to settle and add more slurry until the agarose bed is about 40 μl.

2 Add 1 ml of the affinity buffer and rotate for 5 min to wash. Collect anti-IgG:agarose by centrifugation at 100 g for 1 min. Discard supernatant and repeat the wash. Add 160 μl affinity buffer and 20 μl anti-nucleoprotein antisera to the final pellet. Mix by rotation at RT for 2 h.

3 Collect agarose:anti-IgG:anti-nucleoprotein IgG conjugate by centrifugation at 100 g for 1 min and wash with 1 ml affinity buffer five times as in step 2.

4 Ethanol precipitate cross-linked complexes (*Protocol 36*) isolated from a polyacrylamide gel (*Protocols 13* and *33*) and resuspend in 100 µl complex buffer (fume hood) and add 20 µl 3 M NaCl.[1]

5 Aliquot 3 µl (300–500 c.p.m.) of the complex into separate 0.5-ml microcentrifuge tubes and warm briefly to 60°C. Add 26.7 µl binding buffer, 0.3 µl 1% SDS, 168.3 µl affinity buffer, and 1.7 µl 1% SDS.

6 Add this 200-µl of nucleoprotein complex to the 40-µl bed of agarose:anti-IgG:anti-nucleoprotein IgG conjugate and rotate for 3 h at RT.

7 Collect agarose:anti-IgG:anti-nucleoprotein IgG:nucleoprotein conjugate by centrifugation at 100 g for 1 min and wash with 1 ml affinity buffer twice as in step 2. Measure the activity of the pellet in scintillation counter (^{3}H channel). Positive readings indicate samples have been affinity purified.

8 To release the nucleic acid component, digest the protein with proteinase K, resuspend the pellet in 200 µl protease buffer. Add 200 µl 1 M NaCl and 40 µl proteinase K. Incubate at 37°C for 30 min. Add another 40 µl proteinase K and repeat incubation. Add 2 µl tRNA and 500 µl phenol and shake at 4°C for 45 min on a rotary table.

9 Centrifuge at 5000 g for 10 min in a swing-out rotor and recover the aqueous phase. Ethanol precipitate DNA/RNA and redissolve in a solution appropriate for further experimentation.

Pause points

[1] May be left at –20°C for up to 24 h.

Protocol 37. Antibody affinity isolation of nucleoprotein complexes

Protocol 38. Gel retardation of rRNA:RNA processing complex

Reagents

10 mM Adenosine triphosphate (ATP) (pH 7.5)

Fresh 0.35 mg/ml heparin in diethyl pyrocarbonate (DEPC)-treated water (from 20 mg/ml stock)

Load buffer (95% formamide, 5 mM ethylenediaminetetra-acetic acid (EDTA), 0.03% xylene cyanol, 0.03% bromophenol blue)⚠

Pre-rRNA ^{32}P riboprobe (*Protocol 10*)⚠

10× S-100 processing buffer (0.1 M Hepes-KOH, 0.9 M KCl, 10 mM MgCl$_2$, 1 mM EDTA, 3 mM dithiothreitol (DTT) and 35% glycerol) pH 7.9

S-100 processing extract (ref. 9)

0.5× TBE buffer (4.5 mM Tris-borate, pH 8.3, 1.25 mM EDTA)

Equipment

Autoradiography equipment (cassette, film, developing chemicals)

Gel tank and plates

Water bath at 30°C

Technique (ref. 8)

1 Prepare a nondenaturing 4% polyacrylamide gel (2–2.5 mm thick, 25 cm long, acrylamide:bisacrylamide 65:1) in 0.5× TBE according to *Protocol 18* (urea omitted) and pre-run the gel at 325 V (13 V/cm) for 1 h in 0.5× TBE. Whilst running, carry out steps 2 and 3.

2 Set up the rRNA processing reaction by mixing (on ice): 2.5 μl 5× S-100 processing buffer, 3.75 μl 10 mM ATP, 2 μl (10 fmol) pre-rRNA substrate (riboprobe), 5 μl S-100 processing extract and 11.75 μl water.①

3 Incubate the reaction for 25–45 min at 30°C and place on ice. Add 2 μl

Notes

① This step displaces loosely associated proteins from the rRNA.

0.35 mg/ml heparin solution to the processing reaction but do not mix. Incubate for 10 min at 30°C and place on ice.

4 Load 15 μl reaction mixture on to the pre-run gel (lane 1), 2 μl fresh ribo-probe (control, lane 2), 2 μl riboprobe not exposed to processing extract (but to which steps 2 and 3 have been applied; control, lane 3) and 15 μl load buffer (lane 4).

5 Electrophorese at 13 V/cm until light blue band (xylene cyanol) has travelled the length of the gel.

6 Dry the gel on to filter paper and expose to X-ray film overnight. Judge exposure time and re-expose after developing this film.

Protocol 38. Gel retardation of rRNA:RNA processing complex

Protocol 39. **Gel retardation of DNA:protein complexes**

Reagents

Acrylamide stock solution (deionized 29% acrylamide, 1% bisacrylamide)⚠

10% Ammonium persulfate (APS)⚠ (freshly prepared)

2× binding buffer (40 mM Hepes-NaOH, pH 7.6, 80 mM NaCl, 1 mM dithiothreitol (DTT), 8% Ficoll, 10 mM $MgCl_2$, 0.2 mM ethylenediaminetetra-acetic acid (EDTA))

Nuclear protein (0.5–10 mg/ml)

5′ [^{32}P] end labeled DNA (*Protocol 5*)⚠

Poly(dI-dC) (0.3 mg/ml) in 10 mM Tris-HCl (pH 7.6), 50 mM NaCl, 1 mM EDTA

[^{32}P]pUC19/pBS plasmid (1 mg/ml)⚠

5× TBE (0.45 M Tris, 0.45 M boric acid, 10 mM EDTA, pH 8.3)

N,N,N′,N′-Tetramethylethylenediamine (TEMED)

Equipment

Autoradiography equipment

Electrophoresis apparatus

Heated gel drier

Polyacrylamide gel

SaranWrap

Whatman 3MM paper

Technique

1 Binding reaction series: 10 µl 2× binding buffer, 1 µl [^{32}P] end labeled DNA (5 fmol (20 000 c.p.m.)), 1–9 µl nuclear protein and 9–1 µl poly(dI-dC) in final volume of 20 µl. Leave on ice for 1 h. A negative control of ^{32}P plasmid DNA can be used to replace end labeled DNA (1 µl).①

2 Make a 5% acrylamide gel in 0.2× TBE by mixing 12.5 ml acrylamide stock, 3 ml 5× TBE and 59.5 ml water. Degas solution for 10 min and add 400 µl 10% APS and 30 µl TEMED to initiate polymerization. Pour

Notes

① Poly(dI-dC) optimizes against nonspecific interactions.

solution into prepared 150 × 150 × 2-mm gel plates (*Protocol 16*) and leave at RT to polymerize.

3 Load complex on to the native polyacrylamide gel and electrophorese at 150 V at 4°C for 3 h.

4 Part plates and lay a sheet of Whatman 3MM filter paper over the gel. Pull paper back with gel adhering to the paper. Cover gel with SaranWrap and dry on a gel drier at 80°C.

5 Expose dried gel to X-ray film according to *Protocol 31* for at least 1 day.

Protocol 39. Gel retardation of DNA:protein complexes

Protocol 40. DNA footprinting using DNase I

Reagents

Chloroform:isoamyl alcohol (IAA) (24:1)⚠
10 mM Dithiothreitol (DTT) (fresh)
DNase I (20 ng/μl)
Ethanol⚠
50% Glycerol
10× Hepes buffer (200 mM Hepes-NaOH (pH 7.6), 400 mM NaCl, 20 mM CaCl₂, 50 mM MgCl₂)
Nuclear protein (0.5–10 mg/ml)
5′ [³²P] end-labeled DNA (*Protocol 5*)⚠
Phenol containing 0.1% hydroxyquinoline⚠

Poly(dI-dC) (0.3 mg/ml) in 10 mM Tris-HCl (pH 7.6), 50 mM NaCl, 1 mM ethylenediaminetetra-acetic acid (EDTA)
Sample buffer (95% deionized/fresh formamide, 0.02% xylene cyanol FF, 0.025% bromophenol blue)⚠
Stop buffer (fresh) (2% sodium dodecyl sulfate (SDS), 10 mM EDTA (pH 8), 1 mg/ml tRNA)⚠

Equipment

Gel drier
Sequencing gel apparatus
Water bath at 90°C

Technique

1 Binding reaction: set up a duplicate series of reactions varying protein and poly(dI-dC) concentration, 10 μl 10× Hepes buffer, 10 μl 10 mM DTT, 2 μl (5 fmol/20 000 c.p.m.) DNA fragment, 0–50 μl protein and 50–0 μl of poly(dI-dC), 20 μl 50% glycerol and 8 μl water. Incubate on ice for 1 h.①

2 Add 1 μl DNase I to each reaction and digest one set for 15 sec and one set for 30 sec at RT. Terminate reaction by adding 100 μl stop buffer.

Notes

① Varying poly(dI-dC) concentration optimizes against non-specific interactions.

3 Phenol-extract samples twice with 100 μl phenol and chloroform:IAA extract once. Precipitate with 2.5 vols ice-cold ethanol at −20°C for 30 min. Collect precipitate at 15 000 g for 15 min at 4°C.[1]

4 Dissolve pellet in 10 μl sample buffer and heat denature at 90°C for 2 min. Snap cool on ice prior to electrophoresis. Run sample on a sequencing gel against a sequence of the DNA fragment under analysis (*Protocol 20*).

Pause point

[1] May be left at −20°C for up to 24 h.

Protocol 40. DNA footprinting using DNase I

Protocol 41. Electrophoresis of ribosomes – separation in one dimension

Reagents

30% Acrylamide stock (29% acrylamide, 1% bisacrylamide)⚠
10% Ammonium persulfate (APS)⚠ (freshly prepared)
5 mg/ml Ethidium bromide (EtBr)⚠
Loading solution (10× TBE, 4% Ficoll, 0.01% bromophenol blue)
2-Methylpropan-1-ol⚠
10× TBE (0.9 M Tris, 0.9 M boric acid, 20 mM ethylenediaminete-tra-acetic acid (EDTA), pH 8.3)
N,N,N',N'-tetramethylethylenediamine (TEMED)⚠

Equipment

Electrophoresis equipment
Polyacrylamide gel
Staining tray
UV transilluminator
Video capture apparatus/camera

Technique

1 Make a 5% acrylamide gel in 1× TBE by mixing 12.5 ml acrylamide stock, 7.5 ml 10× TBE and 54.7 ml water. Degas the solution for 10 min and add 200 µl 10% APS and 50 µl TEMED to initiate polymerization. Pour the solution into prepared $150 \times 150 \times 2$-mm gel plates (*Protocol 16*) and leave at RT to polymerize.

2 Place the gel into the tank in a cold room (4°C) and fill the reservoirs with 1× TBE. Leave for 10 min to equilibrate the temperature of the gel. Pre-run the gel at 100 V for 60 min.

3 Load the samples with 1/10 vol. loading solution on to the bottom of the well and electrophorese at 100 V for 5 h.

4 Separate the plates and float the gel off into the electrophoresis buffer containing 5 µg/ml EtBr by squirting the buffer under the gel. Leave for 30 min with gentle agitation.

5 Visualize the RNA bands by placing the gel on a UV transilluminator and record the image either using an image capture device (e.g. Mitsubishi video copy processor) or by Polaroid photography (typical settings: f4.5, 1/30–1/4 sec using Polaroid Polablue (135 mm) film).

Protocol 41. Electrophoresis of ribosomes

Protocol 42. Electrophoresis of ribosomes and nucleosomes – separation in two dimensions

Reagents

30% Acrylamide stock (29% acrylamide, 1% bisacrylamide)⚠
10% Ammonium persulfate (APS)⚠ (freshly prepared)
Equilibration buffer (0.2× TBE, 1% SDS, 20% glycerol, 0.01% bromophenol blue)
Ethidium bromide (EtBr) solution⚠
Loading solution (10× TBE, 4% Ficoll, 0.01% bromophenol blue)
50% Methanol⚠
2-Methylpropan-1-ol⚠
10% Sodium dodecyl sulfate (SDS)
10× TBE (0.9 M Tris, 0.9 M boric acid, 20 mM ethylenediaminete-tra-acetic acid (EDTA), pH 8.3)

1× TBE, 1% SDS
N,N,N',N'-tetramethylethylenediamine (TEMED)⚠

Equipment

Electrophoresis equipment
Polyacrylamide gel
Rotary table mixer
Staining tray
UV transilluminator
Video capture apparatus/camera

Technique

1 Run a native first dimension gel according to *Protocol 41*. Cut the lane of separated nucleoproteins from the gel with a scalpel.

2 Make a 5% denaturing acrylamide gel in 1× TBE by mixing 12.5 ml acrylamide stock, 7.5 ml 10× TBE and 54.7 ml water. Degas the solution for 10 min and add 7.5 ml 10% SDS and 200 µl 10% APS and 50 µl TEMED to initiate polymerization. Pour the solution into prepared 150 ×

Notes

① **Tip:** Place one end of the strip in contact with the gel and slowly push the whole strip into contact with the second dimension gel from this end.

② Polyacrylamide containing SDS cannot be directly stained with EtBr . The methanol dehydrates the gel causing it to shrink. This will be reversed when staining in EtBr.

150 × 2-mm gel plates (*Protocol 16*), leaving a 1.5-cm space at the top and overlayer with 2-methylpropan-1-ol to a depth of 3 mm. Leave at RT to polymerize.

3 Immerse the first dimensional gel strip in the equilibration buffer and shake for 30 min at RT. Carefully decant the buffer off and replace with fresh buffer and shake for 30 min at RT. Place the strip on a piece of tissue to absorb most of the excess buffer.

4 Wash off 2-methylpropan-1-ol from the second dimensional gel and place into the tank. Fill the lower tank reservoir with 1× TBE, 1% SDS. Pipette 1 ml of equilibration buffer on to the top of the gel and load the gel strip on to the gel with a plastic spatula. Fill the spaces between the strips with equilibration buffer and carefully fill the upper reservoir with 1× TBE, 1% SDS.①

5 Electrophorese at 100 V for 2.5 h at RT. Separate the plates with a metal spatula and float the gel off the plate into 50% methanol for 3 h on a rotary table mixer at 10 r.p.m. Change the solution after each hour.

6 Place the gel into 0.1× TBE containing 0.5 µg/ml EtBr. Leave for 30 min with gentle agitation.②

7 Visualize the RNA bands by placing the gel on a UV transilluminator and record the image either using an image capture device (e.g. Mitsubishi video copy processor) or by Polaroid photography (typical settings: f4.5, 1/30–1/4 sec using Polaroid Polablue (135 mm) film).

Protocol 42. Electrophoresis of ribosomes and nucleosomes

Reagents

40% Acrylamide stock (39.4% acrylamide, 0.6% bisacrylamide (60:1))⚠

Agarose

25% Ammonium persulfate (APS)⚠ (freshly prepared)

Electrophoresis buffer (75 mM Tris-glycine (1:13.3 dilution of 1 M stock))

Loading solution (20% glycerol, 0.05% bromophenol blue)

Nuclear extract

N,N,N',N'-tetramethylethylenediamine (TEMED)

1 M Tris-glycine buffer (1 M Tris, 1 M glycine)

Equipment

Hotplate/microwave

Polyacrylamide gel apparatus

Power pack

Technique

1 Make a 3.5%/0.5% composite gel by mixing 5.25 ml acrylamide stock and 24.75 ml water. In a separate Pyrex flask mix 0.8 g agarose, 15 ml 1 M Tris-glycine buffer and 85 ml water. Heat on a hotplate or in a microwave oven to melt the agarose. Cool the solution to 50°C. Transfer 30 ml of the molten agarose to the acrylamide mixture and mix, by swirling immediately.

2 Add 100 µl APS and 100 µl TEMED, mix and pour immediately into pre-pared plates. Leave to polymerize at RT for 1 h.①

3 Place the gel into the electrophoresis tank and fill the reservoirs with electrophoresis buffer. Purge wells by gently squirting electrophoresis

Notes

① Pour gel quickly as a warm solution will polymerize quickly and over-cooling will allow agarose to set.

buffer into the wells. Load the samples without any loading solution as the nuclear extract should be dense. Run a well containing only loading solution alongside to monitor electrophoresis.

4 Electrophorese at 350 V for 5 h, recirculating upper and lower buffers.

5 Separate the plates and electroblot (*Protocol 27*) for hybridization analysis or autoradiograph the gel if labeled complexes were separated.

Protocol 43. Electrophoresis of spliceosomes using composite gels

APPENDIX A: CALCULATION OF MOLES AND MOLARITY

The table below details the number of moles required to make various volumes of solutions of a range of molarities.

Conc (M)	Volume (ml)									
	100	200	300	400	500	600	700	800	900	1000
0.1	0.01	0.02	0.03	0.04	0.5	0.06	0.07	0.08	0.09	0.1
0.2	0.02	0.04	0.06	0.08	0.1	0.12	0.14	0.16	0.18	0.2
0.3	0.03	0.06	0.09	0.12	0.15	0.18	0.21	0.24	0.27	0.3
0.4	0.04	0.08	0.12	0.16	0.2	0.24	0.28	0.32	0.36	0.4
0.5	0.05	0.1	0.15	0.2	0.25	0.3	0.35	0.4	0.45	0.5
0.6	0.06	0.12	0.18	0.24	0.3	0.36	0.42	0.48	0.54	0.6
0.7	0.07	0.14	0.21	0.28	0.35	0.42	0.49	0.56	0.63	0.7
0.8	0.08	0.16	0.24	0.32	0.4	0.48	0.56	0.64	0.72	0.8
0.9	0.09	0.18	0.27	0.36	0.45	0.54	0.63	0.72	0.81	0.9
1.0	0.1	0.2	0.3	0.4	0.5	0.6	0.7	0.8	0.9	1.0

Calculation of moles

One mole of a chemical/molecule is equivalent to its molecular weight (mol. wt) in grams. Therefore, to calculate one mole of an oligonucleotide that is 25 nucleotides (nt) in length:

average nucleotide mol. wt = 310
therefore total mol. wt of oligonucleotide = $310 \times 25 = 7750$
so one mole of oligonucleotide is 7750 g and 1 micromole (μmol) = $(7750 \times 1 \times 10^{-6}) = 7.75$ mg.

Calculation of molarity

Molarity is a measure of the number of moles of a chemical/molecule (e.g. oligonucleotide), in a volume. A molarity of one molar (1 M) is equivalent to one mole (see above) in one liter.

In the laboratory one is usually working with milliliters or microliters of solutions with concentrations as low as picomolar. The table above gives the number of moles needed to make certain volumes of selected molarities. For calculation of millimoles, micromoles, nanomoles and picomoles simply multiply by 10^{-3}, 10^{-6}, 10^{-9} and 10^{-12}, respectively.

DNA and RNA can then be used in gel electrophoresis or labeling studies. Some radiolabeling protocols (e.g. end labeling) rely on a calculation of the number of target sites capable of being labeled. As the target molecules are often of different sizes the protocol procedures address the issue by referring to free ends. The amount of free ends on a molecule may be calculated using the size and concentration of the molecule.

Calculation of free ends

An example to demonstrate this is an oligonucleotide of 25 nt to be end labeled according to *Protocol 7* which requires 20 pmol of free ends. The oligonucleotide has one free end to be labeled which is equivalent to 1/25 of oligonucleotide. Therefore, 20 pmol of free ends is equivalent to 25×20 pmol = 500 pmol of oligonucleotide.

APPENDIX B: SOLUTION RECIPES

Acrylamide stock (30%): 29 g acrylamide (29%) ⚠, 1 g bisacrylamide (1%) ⚠ in 60 ml sterile, double-distilled water. Heat to 30°C until crystals have dissolved. Adjust volume to 100 ml and filter through a 0.45 μm Nalgene filter. Adjust to pH 7 or less and refrigerate until needed.

Acrylamide stock (40%): 38 g acrylamide (38%) ⚠, 2 g bisacrylamide (2%) ⚠ in 6 ml sterile, double-distilled water. Heat to 37°C until crystals have dissolved. Adjust volume to 100 ml and filter through a 0.45 μm Nalgene filter. Adjust to pH 7 or less and refrigerate until needed.

10 M Ammonium acetate: dissolve 77 g ammonium acetate in 80 ml sterile, double-distilled water. Adjust volume to 100 ml and filter sterilize.

10% Ammonium persulfate: Dissolve 1 g ammonium persulfate in 7 ml sterile, double-distilled water before adjusting to a total volume of 10 ml.

1 M Dithiothreitol (DTT): Dissolve 6.18 g of DTT in 40 ml of 10 mM sodium acetate (pH 5.2). Filter sterilize and aliquot. Store at −20°C.

0.5 M EDTA (pH 8.0): Dissolve 186.1 g disodium ethylenediaminetetra-acetate·$2H_2O$ (mol. wt 372) in 800 ml sterile, double-distilled water by adjusting to pH 8.0 with 20.0 g of NaOH pellets ⚠. Make up to 1 liter with sterile, double-distilled water, aliquot and autoclave.

Electrophoresis load buffer: 2.5 ml Glycerol (50%), 0.01 ml 0.5 M EDTA (1 mM), 0.02 g Bromophenol blue (0.04%), 2.49 ml Tris-HCl pH 7.5.

Ethidium bromide (5 mg/ml) ⚠: Add 0.5 g of ethidium bromide ⚠ to 100 ml sterile, double-distilled water and stir on a magnetic stirrer overnight in a foil-wrapped bottle. Store at RT in the light-tight (foil-wrapped) bottle.

GTC solution ⚠: 47.3 g guanidinium thiocyanate ⚠ (4 M), 2.5 ml 1 M sodium citrate pH 7 (2.5 mM), 0.5 g sarcosyl (0.5% w/v) in 100 ml. Autoclave and add 714 µl 14 M 2-mercaptoethanol ⚠ (0.1 M) in fume cabinet prior to use.

2-Mercaptoethanol (0.2 M): 1.4 ml of 2-mercaptoethanol made up to 100 ml with sterile, double-distilled water. Store at 4°C.

1 M MgCl$_2$: Dissolve 101.65 g of MgCl$_2$·6H$_2$O in 400 ml of sterile, double-distilled water. Adjust volume to 500 ml, aliquot and autoclave.

10 × MOPS buffer: 42 g MOPS (0.2 M), 6.8 g sodium acetate (50 mM), 1.9 g EDTA (5 mM) in 1 liter and adjust to pH 7.0 with 10 M NaOH. Store in a dark bottle.

Phenol:chloroform:isoamylalcohol (25:24:1) ⚠: Dissolve 250 g phenol ⚠ in 250 ml chloroform:isoamylalcohol (24:1). Equilibrate the mixture by extracting several times with 0.1 M Tris-HCl (pH 7.6). Store solution under an equal volume of 0.1 M Tris-HCl (pH 7.6) in a dark bottle at 4°C.

10 mM Phenylmethylsulfonyl fluoride (PMSF) ⚠: Dissolve 87 mg of PMSF ⚠ in 50 ml dried isopropanol. Store at 4°C over molecular sieve.

Phosphate-buffered saline (PBS): Dissolve 8 g NaCl, 1.44 g Na$_2$HPO$_4$, 0.2 g KCl and 0.24 g KH$_2$PO$_4$ in 800 ml of sterile, double-distilled water. Adjust to pH 7.4 with HCl and adjust volume to 1 liter. Aliquot and autoclave. Store at RT.

Sepharose CL6B: Supplied by Sigma as a paste. Add equal volume of sterile TE, 0.7 M NaCl and shake to mix. Allow paste to settle and discard wash solution. Repeat wash once. Wash twice as above with T$_{10}$E$_{0.1}$ resuspending the final paste in an equal volume of T$_{10}$E$_{0.1}$, aliquot into 10 ml samples and autoclave. Refrigerate until needed.

Sodium acetate (3 M): Dissolve 204.05 g of sodium acetate·3H$_2$O in 400 ml of sterile, double-distilled water. Adjust pH to 5.2 with glacial acetic acid ⚠ and make up to 1 liter. Aliquot and autoclave. Store at RT.

10% (w/v) sodium dodecyl/lauryl sulfate (SDS): Dissolve 100 g of electrophoresis grade SDS ⚠ in 900 ml of water by heating to 65°C. Adjust pH to 7.2 with conc. HCl ⚠ and aliquot. No need to sterilize, store at RT.

20 × SSC: 175.3 g NaCl (3M), 88.2 g tri-sodium citrate dihydrate (0.3 M) to 1 liter, adjust to pH 7.0 and autoclave.

20 × SSPE: 175.3 g NaCl (3 M), 27.6 g $NaH_2PO_4 \cdot 2H_2O$ (0.177 M), 7.4 g EDTA (0.02 M) to 1 liter and adjust pH to 7.4 with 10 M NaOH ⚠. Aliquot and autoclave.

Tris-HCl pH 7.5 (1 M): Dissolve 121.1 g of Tris base in 800 ml sterile, double-distilled water, allow solution to cool and adjust to pH 7.5 with conc. HCl ⚠; total volume of conc. HCl required is 70 ml. Adjust volume to 1 liter with sterile, double-distilled water.

Tris-acetate (TAE) electrophoresis buffer: 4.84 g Tris base (40 mM), 0.372 g $EDTA \cdot Na_2 \cdot 2H_2O$ (1 mM) dissolved in 800 ml distilled water and pH adjusted to 8.0 with glacial acetic acid ⚠ then made up to 1 liter with distilled water.

Trichloroacetic acid (TCA) (100%): Dissolve 500 g of TCA in 227 ml sterile, double-distilled water.

1 × Tris-phosphate: 0.44 g Tris base (3.6 mM), 0.48 g $NaH_2PO_4 \cdot 2H_2O$ (3.0 mM), 0.04 g $EDTA \cdot Na_2 \cdot 2H_2O$ (0.1 mM), 1.00 g SDS (0.01%(w/v)) (optional) made up to 1 liter with distilled water. Note $NaH_2PO_4 \cdot 2H_2O$ adjusts pH to 7.7 at 25°C.

1 × Tris-borate (TBE): 10.8 g Tris base (89 mM), 5.50 g boric acid (89 mM), 0.93 g $EDTA \cdot Na_2 \cdot 2H_2O$ (2.5 mM) made up to 1 liter with distilled water. Note boric acid adjusts pH to 8.3 at 25°C. Standard buffer used in many molecular biology laboratories.

TE: 1.21 g Tris base (10 mM), 0.372 g $EDTA \cdot Na_2 \cdot 2H_2O$ (1.0 mM), in 1 liter and adjust to pH 7.5 with HCl.

$T_{10}E_{0.1}$:(for DNA sequencing) 1.21 g Tris base (10 mM), 0.037 g $EDTA \cdot Na_2 \cdot 2H_2O$ (0.1 mM), in 1 liter and adjust to pH 7.5 with HCl ⚠.

APPENDIX C: SUPPLIERS

Abbreviation	Company name	Abbreviation	Company name
AIP	Amersham International plc	GRI	Genetic Research Instrumentation Ltd
APP	Appligene	HSI	Hoefer Scientific Instruments
BDH	BDH Laboratories	IBI	International Biotechnologies, Inc.
BHM	Boehringer Mannheim	NBL	Northumbria Biologicals Ltd
BRL	BioRad Laboratories Ltd	PMB	Pharmacia Biosystems
EKL	Eastman Kodak Ltd	PML	Promega Ltd
FIL	Flowgen Instruments Ltd	SCC	Sigma Chemical Corporation
GBL	Gibco–BRL	STG	Stratagene

Amersham International plc., Amersham Place, Little Chalfont, Bucks HP7 9NA. UK.
Tel (0800) 616928, (01494) 544000. Fax (0800) 616927, (01494) 542266.
26236 South Clearbrook Drive, Arlington Heights, IL 60005, USA.
Tel (708) 593 6300. Fax (708) 593 8010.

Anachem Ltd, Anachem House, 20 Charles Street, Luton, Beds LU2 0EB, UK.
Tel (01582) 456666. Fax (01582) 391768.

Appligene, Pinetree Centre, Durham Road, Birtley, Chester-le-Street, Co. Durham DH3 2TD, UK.
Tel (0191) 492 0022. Fax (0191) 492 0617.
1177-C Quarry Lane, Pleasanton, CA 94566, USA.
Tel (510) 462 2232. Fax (510) 462, 6247.

BDH Laboratories, Merck Ltd., Merck House, Poole, Dorset BH15 1TD, UK.
Tel (01202) 669700, (0800) 223344. Fax (01202) 665599.

Beckman Instruments Ltd, Sands Industrial Estate, High Wycombe, Bucks HP12 4JL, UK.
Tel (01494) 441181. Fax (01494) 447558.

BioRad Laboratories Ltd, BioRad House, Maylands Avenue, Hemel Hempstead, Herts HP2 7TD, UK. Tel (01442) 232522, (0800) 181134. Fax (01442) 259118.

Boehringer Mannheim, Bell Lane, Lewes, East Sussex BN7 1LG, UK.
Tel (01273) 480444, (0800) 521578.
9115 Hague Road, PO Box 50414, Indianapolis, IN 46250-0414, USA.
Tel (800) 262 1640. Fax (317) 576 2754.

Clontech & Pharmingen Distributors, Cambridge Biosciences, 25 Signet Court, Newmarket Road, Cambridge CB5 8LA, UK.
Tel (01223) 316855. Fax (01223) 60732.

Clontech Laboratories Inc., 4030 Fabian Way, Palo Alto, CA 94303-4607, USA.
Tel (415) 424 8222. Fax (415) 424 1352.

Calbiochem-Novabiochem Ltd, Boulevard Industrial Park, Padge Road, Beeston, Nottingham NG9 2JR, UK.
Tel (0800) 622935, (0115) 9430840. Fax (0115) 9430951.

Du Pont Ltd, Wedgewood Way, Stevenage, Herts SG1 4QN, UK.
Tel (01438) 734015. Fax (01438) 743621.

Hoefer Scientific Instruments, 654 Minnesota Street, PO Box 77387, San Francisco, CA 94107, USA.
Tel (415) 282 2307, (800) 227 4750. Fax (415) 821 1081.

ICN Flow Biomedicals, Eagle House, Peregrine Business Park, Gomm Road, High Wycombe, Bucks HP13 7DL, UK.
Tel (01494) 443826. Fax (01494) 473162.

International Biotechnologies, Inc., 36 Clifton Road, Cambridge CB1 4ZR, UK.
Tel (01223) 242813. Fax (01223) 243036.

Life Sciences Laboratories, Sedgewick Road, Luton, Beds LU4 9DT, UK.
Tel (01582) 597676. Fax (01582) 581495.

Merck Ltd (BDH) Chemicals, Poole, Dorset BH15 4TD, UK.
Tel (01202) 669700, (0800) 223344. Fax (01202) 665599.

Mitsubishi Cominco, Marunouchi 2-chome, Chioyodaku, Tokyo 100, Japan.
Tel (3) 3213 1321.

Millipore Ltd, The Boulevard, Ascot Road, Croxley Green, Watford WD1 8YW, UK.
Tel (01923) 816375. Fax (01923) 818297.

Eastman Kodak Ltd, PO Box 33, Swallowdale Lane, Hemel Hempstead, Herts, HP2 7EU, UK.
Tel (01442) 42281. Fax (01442) 230367.
25 Science Park, New Haven, CT 06511, USA.
Tel (203) 786 5600, (800) 225 5352. Fax (203) 624 3143, (800) 879 4979.

Flowgen Instruments Ltd, Broad Oak Enterprise Village, Broad Oak Road, Sittingbourne, Kent ME9 8AQ, UK.
Tel (01795) 429737. Fax (01795) 471185.

Fisons Scientific Equipment, Bishop Meadow Road, Loughborough, Leics, LE1 0RG, UK.
Tel (01509) 231166. Fax (01509) 231893.

Genetic Research Instrumentation Ltd, Gene House, Dunmow Road, Felsted, Dunmow, Essex CM6 3LD, UK.
Tel (01371) 821082. Fax (01371) 820131.

Gibco-BRL, Life Technologies, 8400 Helgerman Court, Gaithersburg, MD 20884, USA.
Tel (301) 840 8000.

Hoefer Scientific Instruments, Unit 12, Croft Road Workshops, Croft Road, Newcastle-under-Lyme, Staffs ST5 0TW, UK.
Tel (01782) 617317. Fax (01782) 617346.

NEN Products, Du Pont Ltd, Wedgewood Way, Stevenage, Herts SG1 4QN, UK.
Tel (01438) 734026. Fax (01438) 734379.

New England Biolabs Inc., 67 Knowl Piece, Wilbury Way, Hitchin, Herts SG4 0TY, UK.
Tel (01462) 420616. Fax (01462) 421057.
32 Tozer Road, Beverley, MA 0915-5599, USA.
Tel (508) 927 5054. Fax (508) 921 1350.

Northumbria Biologicals Ltd, Nelson Industrial Estate, Cramlington, Northumberland NE23 9BL, UK.
Tel (01670) 732992. Fax (01670) 732537.

Perkin-Elmer Ltd, Post Office Lane, Beaconsfield, Bucks HP9 2NE, UK.
Tel (01494) 874411. Fax (01494) 679333.
761 Main Avenue, Norwalk, CT 06859-33301, USA.
Tel (203) 762 1000. Fax (203) 762 6000.

Pharmacia Biosystems, 23 Grosvenor Road, St Albans, Herts AL1 3AW, UK.
Tel (01727) 814000. Fax (01727) 814001.
800 Centennial Avenue, PO Box 1327, Piscataway, NJ 08855 1327, USA.
Tel (201) 457 8000. Fax (201) 457 0557.

Promega Ltd, Delta House, Enterprise Road, Chilworth Research Centre, Southampton SO1 7NS, UK.
Tel (01703) 760225, (0800) 378994. Fax (01703) 767014, (0800) 181037.
Madison, Wisconsin, USA.
Tel (800) 356 9526. Fax (608) 273 6967.

Qiagen GmbH, Max-Volmer Strasse 4, 4027 Hilden, Germany.
Tel (02103) 892230. Fax (02103) 892222.
Qiagen Inc., 960 De Soto Avenue, Chatsworth, CA 91311, USA.
Tel (800) 426 8157. Fax (800) 718 2056.

Schleicher & Schuell, 10 Optical Avenue, Keene, NH 03431, USA.
Tel (603) 352 3810. Fax (603) 357 3627.

Sigma Chemical Corporation, Fancy Road, Poole, Dorset BH17 7BR, UK.
Tel (01202) 733114, (0800)447788, Overseas call reverse charge (01202) 733114. Fax (01202) 715460.
PO Box 14508, 3500 DeKalb Street, St Louis, MO 63178, USA.
Tel (800) 848 7791, Overseas call collect (314) 771 5765.

Stratagene, Cambridge Innovation Centre, Cambridge Science Park, Milton Road, Cambridge CB4 4GF, UK.
Tel (01223) 420955, (0800) 585370. Fax (01223) 420234.
11099 North Torrey Pines Road, La Jolla, CA 92037, USA.
Tel (619) 535 5400, (800) 424 5444. Fax (619) 535 0045.

United States Biochemical, PO Box 22400, Cleveland, OH 44122, USA.
Tel (216) 765 5000, (800) 321 9322. Fax (216) 464 5075, (800) 535 0898.

Whatman Scientific, St Leonards Road, 20/20 Maidstone, Kent ME16 0LS, UK.
Tel (01622) 676670. Fax (01622) 677011.

Worthington Biochemical Corporation, Halls Mill Road, Freehold, NJ 07728, USA.
Tel (908) 462 3838, (800) 445 9603. Fax (800) 368 3108, (908) 308 4453.

INDEX